Global Summit on Process Safety 2014

Mumbai, India
15-16 December 2014

ISBN: 978-1-5108-1770-8

Printed from e-media with permission by:

Curran Associates, Inc.
57 Morehouse Lane
Red Hook, NY 12571

Some format issues inherent in the e-media version may also appear in this print version.

Printed by Curran Associates, Inc. (2016)

For permission requests, please contact AIChE
at the address below.

AIChE
120 Wall Street, FL 23
New York, NY 10005-4020

Phone: (800) 242-4363
Fax: (203) 775-5177

www.aiche.org

Additional copies of this publication are available from:

Curran Associates, Inc.
57 Morehouse Lane
Red Hook, NY 12571 USA
Phone: 845-758-0400
Fax: 845-758-2634
Email: curran@proceedings.com
Web: www.proceedings.com

TABLE OF CONTENTS

INVITED SPEAKERS – DAY 1

COMMITTED CULTURE

VIBRANT MANAGEMENT SYSTEMS

RESPONSIBLE COLLABORATION

DISCIPLINED ADHERENCE TO STANDARDS

INTENTIONAL COMPETENCY DEVELOPMENT

INVITED SPEAKERS – DAY 2

Welcome

Scott Berger- CCPS Executive Director

30 Years of Progress Since Bhopal

- CCPS Formed: 1985
- Technical Basis: 1985 – 95
- Management Basis: 1995 – 2005
- CCPS in India: 2004 – Present
- Cultural Basis: 2005 – Present
- Future: Vision 20/20

Looking Forward
Vision 20/20: A Call to Action

- Why? We can now see that goal of zero accidents is achievable
- How will we get there? What is the path?
  - Leverage collective strengths
  - Align efforts
  - Avoid duplication – or worse - contradiction

CCPS GLOBAL SUMMIT ON PROCESS SAFETY

Global Summit on Process Safety

- Help focus drive towards Vision 20/20 and track progress
- Next year: Kuala Lumpur, Malaysia
- Will introduce Vision 20/20 tracking tool
- Continue at least until year 2020

CCPS GLOBAL SUMMIT ON PROCESS SAFETY

Vision 20/20

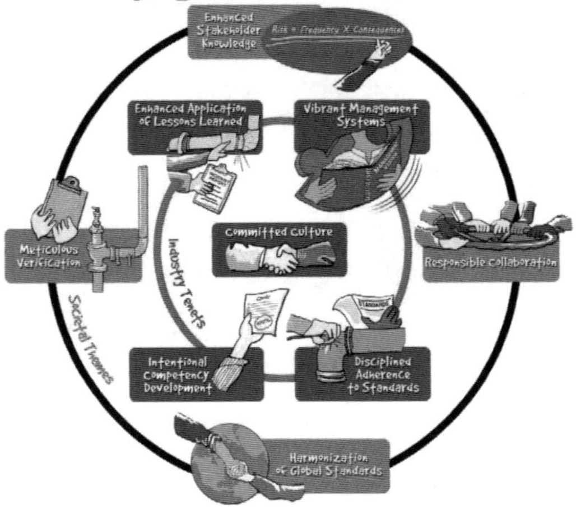

CCPS GLOBAL SUMMIT ON PROCESS SAFETY

Monday 1330

Vision 20/20

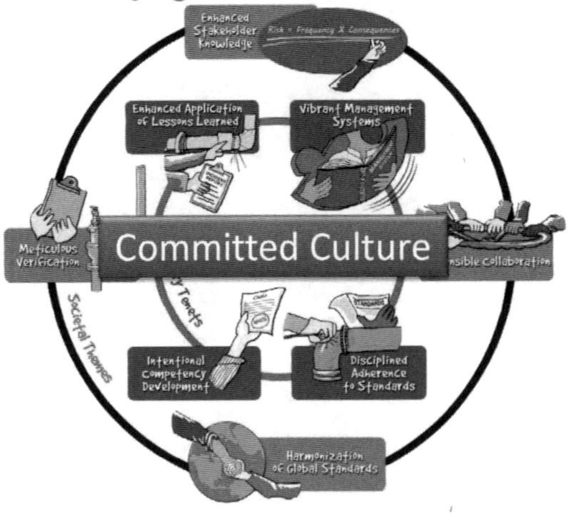

CCPS GLOBAL SUMMIT ON PROCESS SAFETY

Tuesday 830

CCPS GLOBAL SUMMIT ON PROCESS SAFETY

Monday 1330

CCPS GLOBAL SUMMIT ON PROCESS SAFETY

Monday 1600

CCPS GLOBAL SUMMIT ON PROCESS SAFETY

Monday 1600

CCPS GLOBAL SUMMIT ON PROCESS SAFETY

Tuesday 1100

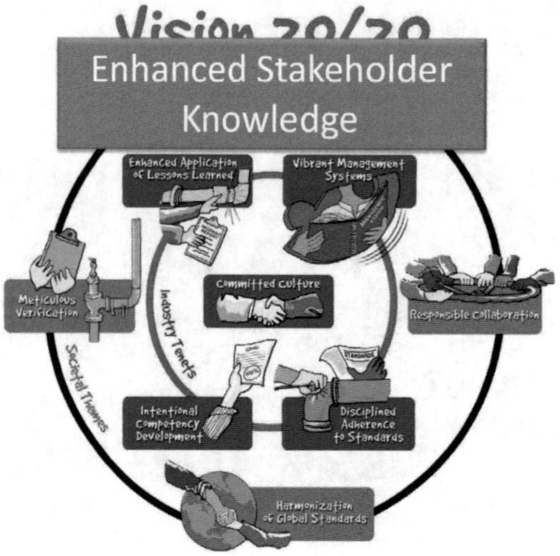

CCPS GLOBAL SUMMIT ON PROCESS SAFETY

Tuesday 1400, led by UNEP

CCPS GLOBAL SUMMIT ON PROCESS SAFETY

Tuesday 830

CCPS GLOBAL SUMMIT ON PROCESS SAFETY

Tuesday 1100

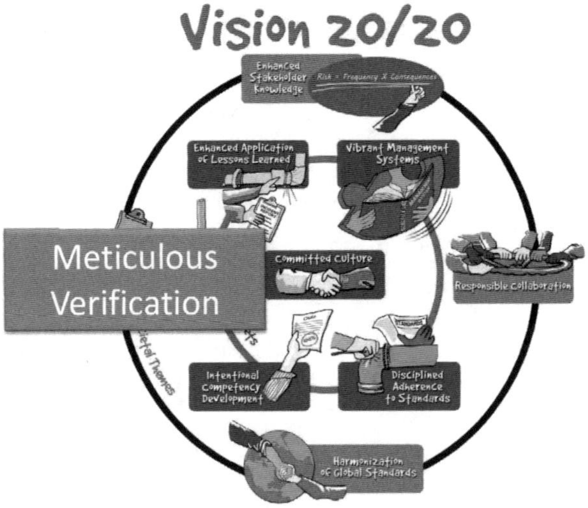

CCPS GLOBAL SUMMIT ON PROCESS SAFETY

Leadership Forum

- Invitees: Senior business and operations executives
- "How I drive process safety in my organization"
- "How I am positioning my organization for Vision 20/20"

CCPS GLOBAL SUMMIT ON PROCESS SAFETY

Your job this week

- Share your experiences: We must learn from each other and learning cannot go just one way
- Find the gems: Take particular note of things you hear which you can use immediately
- Take action: Identify opportunities which you will implement when you return to work
- Expand your network: Build relationships with CCPS staff and other attendees

CCPS GLOBAL SUMMIT ON PROCESS SAFETY

CCPS GLOBAL SUMMIT ON PROCESS SAFETY

An AIChE Technology Alliance
CCPS
Center for Chemical Process Safety

Case Study: Laser-based gas detection technology and dispersion modeling used to eliminate false alarms and improve safety performance on Terra Nova FPSO

Edward Sharpe
Suncor Energy, Inc.
esharpe@suncor.com

Rajat Barua
Senscient, Inc.
rbarua@senscient.com

Are Bratteteig
GexCon US
are@gexcon.com

Zona Bath
Suncor Energy, Inc.
zbath@suncor.com

Abstract

Suncor Energy is the operator of the Terra Nova FPSO, which is located in the Grand Banks, off the East Coast of Canada. In 2010, a multi-disciplinary team was assembled to assess and upgrade the overall gas detection system on the FPSO. The team involved personnel from safety, risk analysis, operations, instrumentation and controls engineering. A detailed analysis of the facility, based on computational fluid dynamics (CFD) modeling, was performed. In aggregate, more than 1,400 gas leak scenarios were simulated and used in the evaluation, detector selection process, optimization and overall design of the upgrade to the gas detection system. Laser-based technology was selected to replace infrared gas detection technology after extensive testing in both onshore and offshore environments. By combining a quantitative gas dispersion study with the implementation of new technology, Terra Nova was able to achieve:

1. An elimination of false alarms
2. An increase in gas leak detection coverage
3. Earlier warning for preventative and remedial action
4. A reduction in maintenance requirement
5. A reduction in the exposure of operations personnel to hazardous locations and gases
6. An improvement in the reliability and robustness of the FPSO's overall gas detection system.

The Terra Nova FPSO is a remote facility with limited egress. Therefore, any hazardous gas release in the facility requires complete production shutdown, blow down of available inventory and isolation of electrical equipment that is not Zone 1 rated. Prior to the upgrade, false alarms from gas detectors were resulting in prolonged outages, damage to process equipment and production deferments of approximately 50,000-100,000 barrels per year. The upgrade was able to address all of these problems. This paper describes the methodology that was applied and presents an overview of the results. Implementation of this retrofit and upgrade approach is expected to benefit numerous industrial facilities where the threat of a toxic or flammable gas leak exists.

Introduction and background

The Terra Nova oil field is situated in the Grand Banks, approximately 350 km (220 miles) east-southeast of the city of St. John's in the Canadian province of Newfoundland and Labrador. The field is the second largest producing oil field in East Coast Canada and is located in the Jeanne d'Arc basin (see figure 1). The total recoverable oil reserves in the field are estimated by the Canada-Newfoundland Labrador Offshore Petroleum Board (C-NLOPB) to be 419 million barrels. The owners of the Terra Nova field are: Suncor Energy (37.675%), ExxonMobil (19%), Statoil (15%), Husky Energy (13%), Murphy Oil (10.475%), Mosbacher Operating Ltd. (3.85%), and Chevron Canada (1%). Suncor Energy is the operator of the Terra Nova field.

Discovered in 1984, the field was the second to be developed off the coast of Newfoundland and Labrador. Production from the field began in 2002, through the use of a Floating, Production Storage and Offloading (FPSO) vessel. This was the first development in North America to use FPSO technology. One of the largest FPSO vessels ever built, the Terra Nova FPSO is 292.2 meters long and 45.5 meters wide, approximately the size of three football fields laid end to end. From the keel to the helideck, it stands more than 18 stories high (see figure 2). The Terra Nova FPSO can store 960,000 barrels of oil and accommodate up to 120 personnel while producing. Oil production wells are pre-drilled by a semi-submersible mobile offshore drilling unit. The wellheads and production manifolds are placed in “glory holes” (excavations in the seafloor) that protect the equipment from potential scouring by icebergs. A network of more than 40 kilometers of flexible flowlines is used to convey hydrocarbons to and from the wells. Produced gases are separated from the oil and re-injected into the reservoir for production support and possible future extraction. Crude oil is offloaded from the FPSO onto large shuttle tankers for shipment.

Environmental conditions and challenges

The Grand Banks region has a harsh environment, similar to that of the northern North Sea. The FPSO is situated in "iceberg alley" where large icebergs from Greenland and Ellesmere Island drift south with ocean currents. The Terra Nova FPSO is a double-hulled, ice-reinforced vessel, with five thrusters and a dynamic positioning system, which enables the FPSO to change to more favorable headings in high winds and storms. The cold Labrador Current from the north and the much warmer Gulf Stream from the south collide in this area to produce very challenging weather conditions (see figure 3). In June and July, the FPSO can be covered in thick fog 84% of the time and, with over 200 foggy days each year this is one of the foggiest places in the world. During the fall and winter, the Terra Nova FPSO can experience intense storms whereby wave heights exceed 25 meters and wind speeds exceed 100 knots. Snow squalls, freezing rain and freezing ocean spray occur frequently in the winter months.

Gas detection history

The original Terra Nova gas detection system was designed to utilize a combination of point detectors and line-of-sight (LOS) detectors based on infrared (IR) technology, with the bulk of coverage being provided by the LOS type. Since the facility first started production in 2002 there have been ongoing issues with the reliability of the LOS detectors. The most significant issue has been performance during adverse weather conditions. During periods of fog or snow many detectors would report a blocked beam fault condition and, depending on the situation, some would falsely detect gas. As a result, Suncor would frequently be forced to assign technicians to clean the lenses and return the units to service. Simultaneously, operations staff would also place inhibits on many detectors to prevent inadvertent plant trips. In any zone, once a threshold of 50% of detectors requiring inhibits had been crossed, temporary portable detectors were deployed and operating personnel would need to be present more frequently in the affected areas to monitor for gas releases.

As the Terra Nova FPSO is a remote facility with limited egress, any gas release in the facility requires a complete production shutdown, blow down of available inventory and isolation of electrical equipment that is not Zone 1 rated. To achieve this, the FPSO was designed with multiple fire and gas detection zones. Each of the zones utilizes one of two voting methodologies depending on coverage and detectable gas cloud size. The first and most widely used method is 1oo2D. This method is a typical 2ooN voting system with the addition of diagnostic coverage, which simply means than any detector that is in fault counts as a vote. For example, a zone with 5 detectors would require 2oo5 for executive action to occur. If one detector enters into a fault condition the zone then becomes 1oo4. The second methodology is 1ooN. This voting methodology is limited to the main vessel deck area of the FPSO and was selected due to the low number of detectors, the confinement in the area and the need to detect small clouds. Many of the false plant trips have occurred due to detectors in the 1ooN zones.

Advent of sour gas

By 2010 parts of the Terra Nova field were beginning to produce hydrogen sulfide (H_2S) in the well fluids. As toxic gases were not expected in the original Terra Nova design, the fixed gas detection system did not incorporate dedicated H_2S detection. Based on the need to upgrade the gas detection system to detect H_2S and the LOS detector performance issues, several projects were initiated. The first initiative was to build a gas dispersion model for the Terra Nova FPSO and use this analysis to complete a comprehensive gas detection evaluation and optimization study. In parallel, Suncor began to investigate and trial various gas detection technologies that could address the problems with the existing system and detect both flammable and toxic gas releases.

Evaluation of the gas detection system

Detailed analysis based on computational fluid dynamics (CFD) modeling was performed for the Terra Nova FPSO by GexCon. The overall objective of the analysis was to test and optimize system performance and suggest improvements for the existing system and detector layout. The process included:

1. Accurately representing ventilation conditions across all process modules
2. Simulating a range of realistic gas leak cases
3. Simulating explosions to establish dangerous cloud sizes for each module
4. Suggesting and benchmarking a range of detector designs and layouts for each module
5. Arriving at a recommended optimum system for each module, accounting for limitations on number and type of detectors, input / output limitations, and voting schemes
6. Ensuring sufficient performance for both toxic and hydrocarbon gas detection

In aggregate, more than 1,400 gas leak scenarios were simulated and used in the evaluation, optimization, detector selection and overall design of the upgrade to the gas detection system.

The CFD model FLACS was chosen in the present study because it allows (1) the simulation of the relevant toxic and flammable gaseous releases, (2) the modeling of the detailed geometry within the facility, (3) the simulation of both natural and forced ventilation within the facility, (4) the evaluation of gas concentration measured by line and point detectors, (5) the evaluation of danger potential for the dispersion cloud, and (6) the ability to perform parametric sensitivity and optimization studies. The geometry of the Terra Nova FPSO was based on a computer-aided design (CAD) model. In order to produce reliable results from the simulation, it was of great importance that the geometry model be sufficiently accurate. Obstructions will restrict the air flow and result in reduced ventilation rates, diversion of flow, or both. Having a geometry model with representative congestion and confinement characteristics assures the reliability of both the ventilation and dispersion simulations.

Gas dispersion simulations

In order to assess the performance of the gas detection system (GDS) with regards to safety, the following criteria were used:

1. A confirmed gas alarm must be triggered before a leak reaches a hazardous level. This hazardous level occurs when the cloud of gas from the release becomes larger than the 'dimensioning cloud size'. This dimensioning cloud size is the smallest flammable cloud, which, if ignited, could break through walls or decks or otherwise cause escalation and is determined by performing explosion simulations for a range of cloud sizes. As an example, the V30 module covers the entire main deck of the FPSO. Explosion simulations were performed to identify the size of cloud that could exceed the design strength of the main barriers. The congested and confined nature of the V30 area meant that relatively small flammable gas clouds could threaten the integrity of the deck, resulting in a dimensioning cloud size of $88m^3$. The dimensioning cloud sizes for other main areas and modules (refer

to figure 4) of the FPSO were determined as follows:

- Turret Area (T06) – 1,585m^3
- Water Injection Module (M02) – 604m^3
- Separation and High Power Compression Module (M03) – 556m^3
- Produced Water / Glycol Module (M04) – 406m^3
- Separation and Low Power Compression Module (M05) – 332m^3
- Power Generation Module (M09) – 695m^3

2. Low alarm must be triggered for any leak scenario where the volume with an H_2S concentration above 10ppm exceeds 100m^3.
3. Low alarm must be triggered within 30 seconds for any leak scenario where the volume with an H_2S concentration above 100ppm exceeds 10m^3. The concentration level of 100ppm was selected because it is defined as being Immediately Dangerous to Life and Health (IDLH) by the US National Institute for Occupational Safety and Health.

Beyond these three key criteria, it was decided that the system should be optimized to give an alarm as early as possible. Judging performance by time to detection alone would miss the spatial extent of the gas cloud.

The quantitative evaluation of the GDS in the process modules was based on testing the performance for a large set of realistic gas leak scenarios. This means that ventilation conditions and gas leak properties had to be modeled as accurately as possible. The local ventilation conditions were critical in understanding the formation and migration of flammable or toxic gas clouds. The ventilation conditions were determined by the wind speed and direction, the heading of the FPSO, large-scale effects from modules and buildings upwind and downwind of the area of interest, as well as smaller-scale effects from buildings, walls, equipment and piping inside the area of interest itself. An aerial view of the entire FLACS model of the FPSO is shown in figure 5.

Gas detector technology selection

Since it commenced operation in 2002, the Terra Nova FPSO has relied primarily on LOS detection. Therefore, Suncor's preferred option was to determine if LOS technology existed that could replace its existing detectors and provide coverage for both flammable and toxic gas leaks. This would minimize structural modifications required to mount new units, avoid having to run new cables and trays, thereby minimizing construction time and cost. An extensive search eventually led the Suncor team to lab test, field trial and ultimately select the Enhanced Laser Diode Spectroscopy (ELDS™) technology. This technology, developed by Senscient, had the capability of combining toxic and flammable gas detection in a single LOS detector whilst significantly improving the performance and reliability of the FPSO's overall gas detection system.

In February 2011, a trial ELDS™ unit was acquired and subjected to qualification testing at the

Terra Nova onshore distributed control system (DCS) simulator. Over a period of two weeks, stringent performance tests were conducted on the device, under a variety of conditions that were known to generate problems for the existing IR detectors. The test results are summarized below.

Test Description	Result
Simulated fog using water mist	No issue with performance
Direct water spray on the device lenses	No issue with performance
Plastic of various types placed in the beam path	No issue with performance
Snow placed over the lens (approx. 1 inch thick)	No issue with performance

Using a transmitter-receiver configuration, ELDS™ systems detect and measure gas concentrations at specific target gas absorption wavelengths over distances of up to 200 meters (see figures 6 and 7). The transmitter utilizes a highly reliable, solid-state laser diode sources similar to those used in demanding telecommunications applications to generate a laser beam. The receiver measures absorbance changes when a combustible or toxic gas passes through the laser beam. One of the key techniques that enables ELDS™ to detect small fractional absorbances and eliminate false alarms is Harmonic Fingerprinting. A Harmonic Fingerprint is a specific set of harmonic components introduced by target gas absorption where the relative amplitudes and phases of the components are known and specific to the target gas absorption line that is being scanned (see figure 8 for an example).

Using a small retained sample of target gas inside the transmitter, the temperature and wavelength modulation currents applied to the transmitter's laser diodes are actively controlled to lock the lasers such that absorption by target gas produces specific Harmonic Fingerprints. The relative amplitudes and phases of the harmonic components in a Harmonic Fingerprint are so specific and unique that only absorption by the specified target gas produces a signal with the desired Harmonic Fingerprint. Noise, absorption by atmospheric gases and coherent interference effects never produce signals with the Harmonic Fingerprint, enabling an ELDS™ based gas detector to eliminate false alarms (see figure 9). In addition to maintaining a Harmonic Fingerprint lock, the gas reference cell enables remote, on command, electronic functional testing of the gas detector either locally or from the control room under any condition. In the test, the Transmitter simulates a predefined quantity of target gas by directing the laser beam through the reference cell as seen in figure 12. ELDS™ units are programmed to conduct a validity test every 24 hours. The results of detector functionality testing are gathered and logged automatically. The presence of a gas reference cell is an innovation that improves reliability and reduces maintenance, which eliminates the need for technicians to carry cylinders of hazardous gases through the FPSO in order to test gas detectors.

Optimization of the gas detection system

In order to verify and optimize the performance of the GDS in each module, a diverse and challenging set of leak scenarios was needed. Leak positions were selected that resulted in flammable and toxic clouds in all parts of the module, so that the entire GDS could be evaluated quantitatively. Leak rates that yielded challenging clouds were chosen, i.e. large enough to be dangerous, but small enough to be difficult to detect (see figure 10 for an example). The gas detectors on the vessel are logically split into fire zones. Where 1oo2D or 2ooN voting is used, both contributing detectors must be in the same fire zone in order for the alarm to be triggered. This voting logic was fully implemented in the CFD testing of detector layouts. Though the fire zone splits are mostly by location, in some cases, detectors located physically close to each other are logically placed in different fire zones.

For every module, detection results were summarized and presented as graphs showing cloud size at detection for simulated scenarios. This format shows the performance of a GDS at a glance, but requires a closer description. Figure 11 shows an example graph prepared for detection of hydrocarbon gas clouds. Detected scenarios are ordered by cloud size at detection (increasing order). Undetected scenarios are shown after the detected scenarios, in descending order by maximum cloud size. To discern between detected and undetected scenarios, the latter are shown with negative values (i.e. the lower the value, the larger the undetected cloud). Also shown are the threshold values – the dimensioning cloud size for hydrocarbon detection, $100m^3$ or $10m^3$ for H_2S detection (these threshold values correspond to the three criteria listed previously). A curve that dips below the lower red line represents undetected scenarios over the threshold size, which means the GDS does not satisfy the detection criterion. A curve that crosses the upper red line represents scenarios that are detected, but only after the cloud reaches threshold size. For hydrocarbon detection, this means the GDS does not satisfy the detection criterion, while the H_2S detection criteria are not directly tied to cloud size at detection.

Presenting results for the entire FPSO would be an extremely lengthy endeavor, well beyond the scope and space limitations of this paper. Instead, it is possible to illustrate the methodology and results by focusing on one module. The M02 module, comprising of the water injecting area, can be considered as such an example. Explosion simulations were performed to identify the size of cloud that could exceed the design strength of the main barriers and this analysis resulted in a dimensioning cloud size of $604m^3$. Gas dispersion simulations were then performed with a range of different release locations. The majority of the releases were placed on the process deck as this area contains most of the potential leak sources. The confinement and flow pattern in this area generated a wide range of cloud shapes. In aggregate, 253 gas dispersion simulations were performed for module M02. Following this, two optimization rounds were performed in an iterative manner. The first round was used to test the sensitivity of the system. The second run focused on finding the optimal layout. Figure 12 shows performance graphs for several tested GDS solutions in the water injection area (module M02).

- The original GDS (black curve titled "BC") has the worst performance, signified by the early drop-off (signifying many scenarios are undetected) and the deep trough (showing that the undetected scenarios are relatively large).
- The first attempted improvement was simply to combine voting across fire zones. The grey-blue "Combined Firezones" shows a slight improvement – a lower peak, and slightly later drop-off signifying a few more scenarios are detected. Both this and the BC curve dip below the lower red line, and thus do not satisfy the performance criterion.
- Next, the effect of the planned transition to laser-based LOS technology, combined with the aforementioned combined voting was evaluated. The green "Combined Senscient" has a much later drop-off, detecting many scenarios that are not detected by the two previous curves. However, it crosses the upper red line where those curves did not. This may seem counter-intuitive, but is actually a sign of improved performance. Some scenarios that were previously undetected are detected, even though the detection occurs after the cloud has reached a large size. While this represents an improvement over not detecting these scenarios, this curve still does not satisfy the criterion.
- Further, several alternative layouts with additional detectors were evaluated. The brown "L2 Senscient" curve shows further improvement, with a lower peak, later drop-off and shallower trough. As it crosses both of the red lines, it is still not in compliance with the performance criterion.
- Finally, another tested GDS, the blue "L3 Senscient" shows the best performance of these curves. While it still has a drop-off (signifying undetected scenarios), the undetected scenarios are smaller than the dimensioning size.

The GDS solution illustrated by the blue curve entitled "L3 Senscient" was selected for implementation. With this solution, all gas leak scenarios bigger than $604 m^3$ (the dimensioning cloud size for module M02) are detected before reaching that size. The curve does not cross any of the red lines and satisfies the criterion for hydrocarbon detection.

Implementation of the retrofit and upgrade

Once onshore testing was completed, a dual Methane-H_2S ELDS™ detector was installed on the Terra Nova FPSO, in May 2011, in place of an existing IR LOS device that was known to cause issues in the past. The distance between the transmitter (Tx) and receiver (Rx) for this installation was approximately 34m. During a trial period that lasted approximately 24 months, the ELDS™ detector performed very well, with zero spurious trips. The only issue identified was a loose mounting bracket that resulted in misalignment, a problem that was easy to identify and resolve. Following the successful offshore trial, a decision was made by Suncor to replace all existing IR LOS gas detectors on the Terra Nova FPSO with the ELDS™ technology. In aggregate, 141 IR LOS detectors were slated for replacement. A recommendation from the study by GexCon added another 17 ELDS™ detectors to provide for additional coverage. Therefore, the overall retrofit and upgrade plan called for the decommissioning or 141 IR LOS and the installation of 158 ELDS™ detectors.

Retrofitting the FPSO while it was on location, in full operation, required significant planning. A thorough work package was developed which provided a plan for the decommissioning of the existing IR detectors, installation of the new ELDS™ detectors and training of personnel. Change out of the LOS detectors commenced in 2013. As of July 2014, 78 of 158 ELDS™ detectors have been installed, with the remaining devices scheduled to be changed through the remainder of 2014 and 2015.

Results and discussion

The performance of the upgraded system has been exceptional. Prior to the upgrade, the Terra Nova FPSO experienced 3-5 plant trips, more than 100,000 fault indications and 20-25 unrevealed failures per year. Data from the Terra Nova data historian, maintenance management system and lost production tracking register was analyzed to quantify this performance. The sections below summarize the improvement in performance to date.

Impact on safety performance

A review of the maintenance management system showed that there was an average of 21 unrevealed gas detector failures per year from 2009 to 2013. The signal these IR devices were sending to the control system indicated full health, however, they failed to respond to a function check. The ELDS™ detectors address this problem by completing a daily diagnostic check and signaling a fault if there is an issue. As of July 31, 2014, Terra Nova had not recorded a single unrevealed failure with the laser-based gas detectors. Startup and shutdown of a facility introduce risks that are not present when a plant is in steady state operations. By reducing the number of spurious trips, the time spent in startup or shutdown modes is reduced, thereby decreasing the risk to the facility.

The laser-based sensors installed on the Terra Nova FPSO have a minimum detection threshold that is much lower than the older IR-based detectors. Instead of a hydrocarbon detection threshold of 1 LELm / 2.5 LELm (low / high alarm), the laser-based detectors can be set to a threshold of 0.2 LELm/1 LELm (low / high alarm), without experiencing any drifting and related spurious alarms. The effect of this five-fold increase in sensitivity is that the detectable volume of any gas cloud is larger. Figure 14 shows two plots of the same simulated gas leak. The red zone is flammable gas, whereas the yellow zone represents detectable volume with a sensitivity of 10% LEL and 50% LEL, respectively for the top and bottom plots. Increasing the sensitivity of detectors increases the detectable volume and makes it more likely that any given detector will be exposed to detectable gas before the flammable volume reaches a dangerous size (see figure 15). Furthermore, the H_2S content on Terra Nova means even small leaks can be hazardous, thus increased sensitivity is an important tool to ensure worker safety.

In some areas of the FPSO, detectors that were physically close belonged to different fire zones. Since voting is done on a per-fire zone basis, this organization could lead to cases where several detectors must be exposed before a confirmed alarm is achieved. Results obtained from the gas dispersion model showed that, for some modules, combining voting across neighboring fire zones significantly improved detection performance even in the absence of any other improvement. For example, in modules M03, M04 and M05, voting across the fire zones in each module is being combined and this increases the percentage of scenarios detected before reaching dangerous size to 100% from 98%, 99% and 91%, respectively.

A comprehensive dispersion analysis based on worst case H_2S concentrations in the production stream was performed for the entire FPSO. This study revealed that hazardous levels of H_2S could occur in the port half of module M04, the area of the process facility where regeneration of glycol and treatment of gas and produced water takes place. For other areas of the FPSO, results obtained from the gas dispersion modelling demonstrated that the H_2S concentrations were low and thus detection was adequately ensured via the optimized hydrocarbon GDS. Consequently, a dedicated H_2S detection system was developed and installed in the M04 module utilizing laser-based LOS devices. The installed system satisfied the performance criteria for H_2S detection.

Impact on maintenance and offshore operations

A review of performance data from the Terra Nova data historian shows dramatic improvement over the IR devices in spurious trip avoidance. The key metric reviewed was the number of faults / blocked beam indications. Performance data was reviewed from January to December of 2011 for the IR-based detectors as the FPSO was off station for maintenance for portions of 2012 and 2013. Data from June 2013 to June 2014 was used to capture the performance of detectors that had been changed to laser-based systems. As the number of devices and time in service for the two groups is not equal, a metric of 'fault rate per detector day in service' is used for comparison purposes (total faults / sum of days in service). A summary of the data is provided below.

	Faults (total)	Faults / Day
Infrared (IR) LOS Detectors	206,323	3.82
Laser (ELDS^{TM}) LOS Detectors	1,899	0.14

It should be noted that in both cases there were a large number of faults attributed to detectors chattering in and out of a fault condition. The table above includes all faults regardless of the cause. In the case of the laser-based detectors 1,561 of the 1,899 faults were attributed to two detectors that had alignment issues over several days before being corrected. Outside of these two detectors, the fault rate per detector was approximately 0.025 faults per detector per day and the vast majority were due to alignment issues. For the IR detectors, one device accounted for nearly half of the faults. This IR device had a failed source that resulted in a chattering fault for a six-month period while waiting on parts for replacement. Overall, the fault rate, not counting the problem detector, was significantly higher for IR than for ELDS™, with 1.1 faults per detector per day with the majority due to weather, dirty optics and alignment.

With such a significant decrease in the number of faults, the requirement for technician response and troubleshooting has been drastically reduced. From 2009 to 2013, an average of 234 maintenance work orders per year was generated, relating to IR LOS detectors being in fault. Note that this number does not include the significant number of maintenance call outs during conditions of limited visibility (rain, fog, mist and snow) to clean detectors as these calls are not tracked. From June 2013 to July 2014, there were only 13 maintenance work orders created relating to issues with laser-based devices.

Impact on production

Historically, the Terra Nova FPSO has experienced three to four production trips per year due to false gas detection. These trips have resulted in production deferments of approximately 50,000-100,000 barrels per year. During a trip initiated by gas detection, emergency shutdown valves are closed quickly, equipment is tripped and gas inventory is sent to the flare system. Shutting down the plant in this manner has potential damaging effects on plant equipment such as generators and gas compressors. The FPSO experienced damage to process equipment during emergency shutdowns several times prior to the installation of the laser-based gas detectors, resulting in prolonged outages and significant repair costs. Conversely, there has not been a single instance of false gas detection with the laser-based gas detection technology.

Conclusion

This paper presents the methodology that Suncor adopted to tackle problems that were impacting safety, maintenance and production operations. The advent of sour gas in the production fluid stream and detector failures, in particular the problem of false alarms due to adverse weather, were the catalyst for an extensive overhaul and upgrade of the gas detection system on the Terra Nova FPSO. A detector optimization study based on computational fluid dynamics, combined with the installation of advanced laser-based gas detection technology, has solved these problems and resulted in a significant improvement in safety, reliability and process uptime. Implementation of this approach is expected to benefit numerous industrial facilities where the threat of a toxic or flammable gas leak exists.

References

1. Skaug, M., Kerwin, K., & Katay, J. (2001, January 1). Reservoir Development Plan for the Terra Nova Field. Offshore Technology Conference. doi:10.4043/13024-MS
2. Haugen, E. D., Costello, J. P., Wilcox, L. B., Albrechtsons, E., & Kelly, I. (2007, January 1). Reservoir Management Challenges of the Terra Nova Offshore Field: Lessons Learned

After 5 Years of Production. Society of Petroleum Engineers. doi:10.2118/109587-MS

Figure 1. Location of the Terra Nova field in the Jeanne d'Arc basin.

Figure 2. The Terra Nova FPSO.

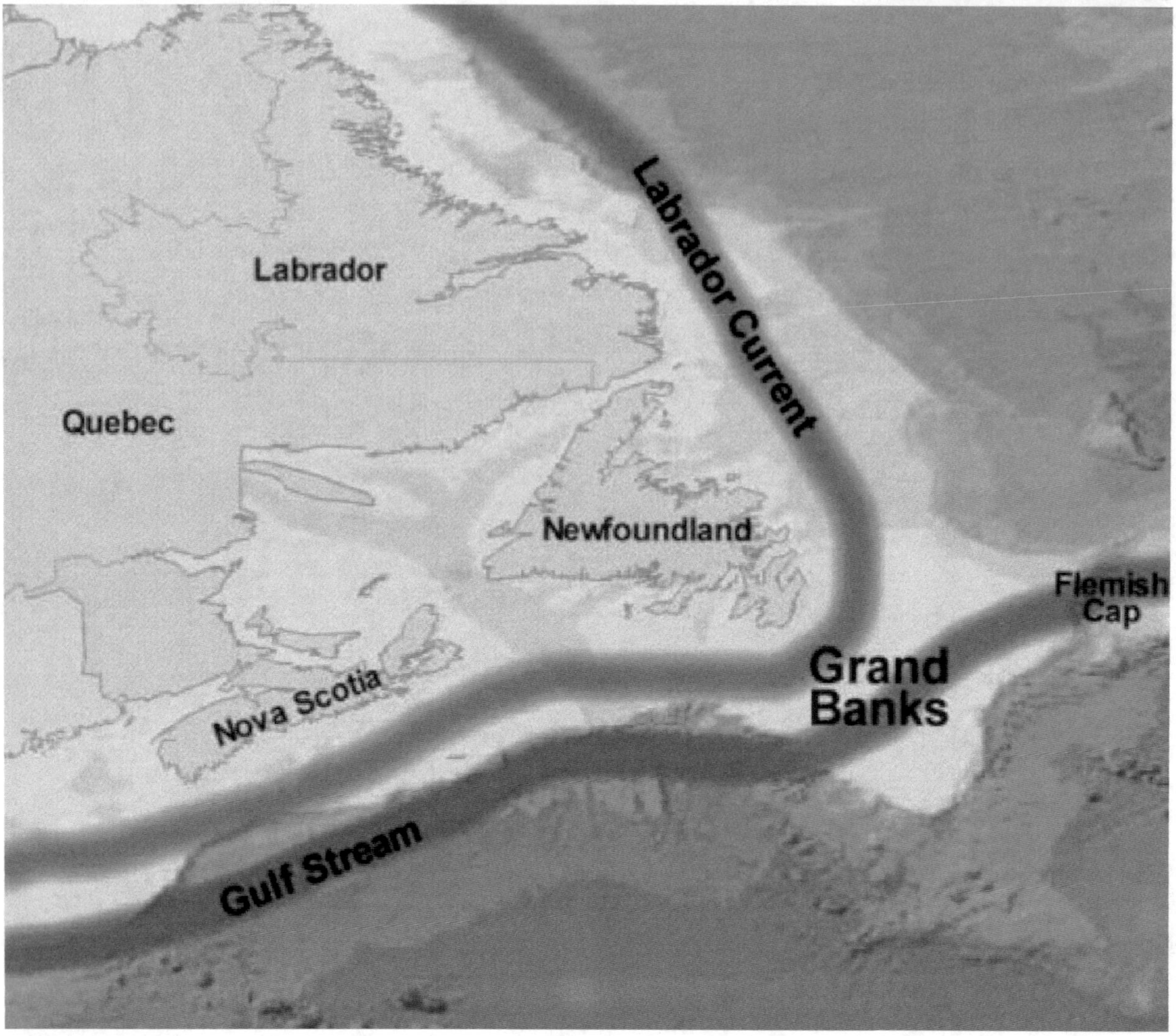

Figure 3. The cold Labrador Current from the north and the much warmer Gulf Stream from the south collide in the Grand Banks to produce very challenging weather conditions.

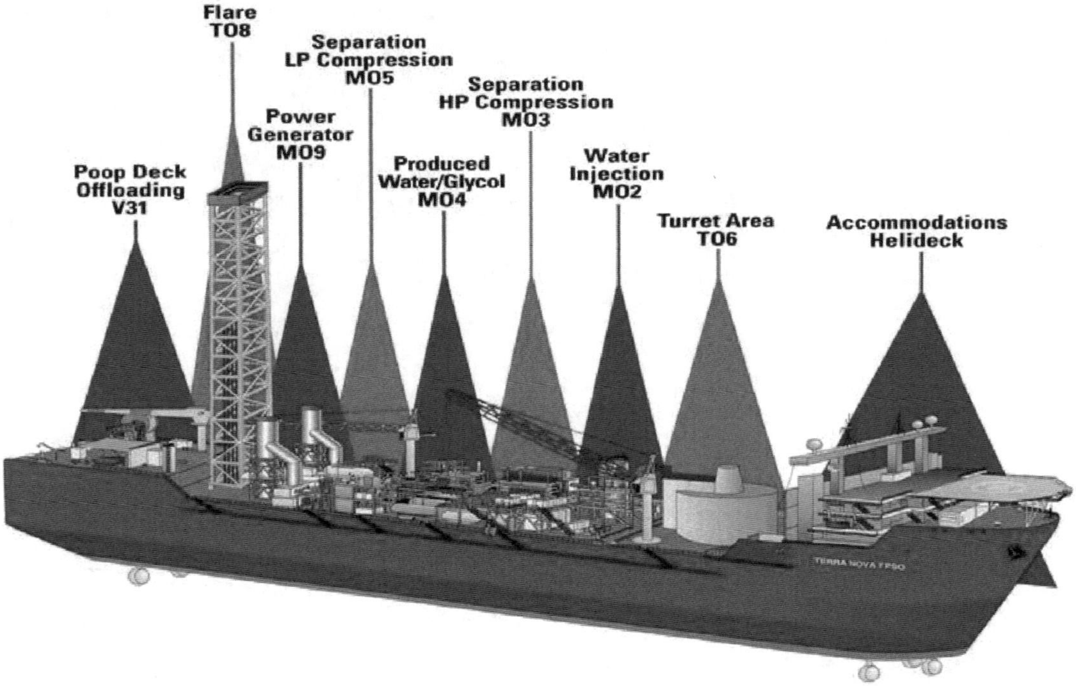

Figure 4. A graphical illustration of the Terra Nova FPSO that indicates key areas / modules.

Figure 5. A graphical illustration of the FLACS geometry model of the Terra Nova FPSO.

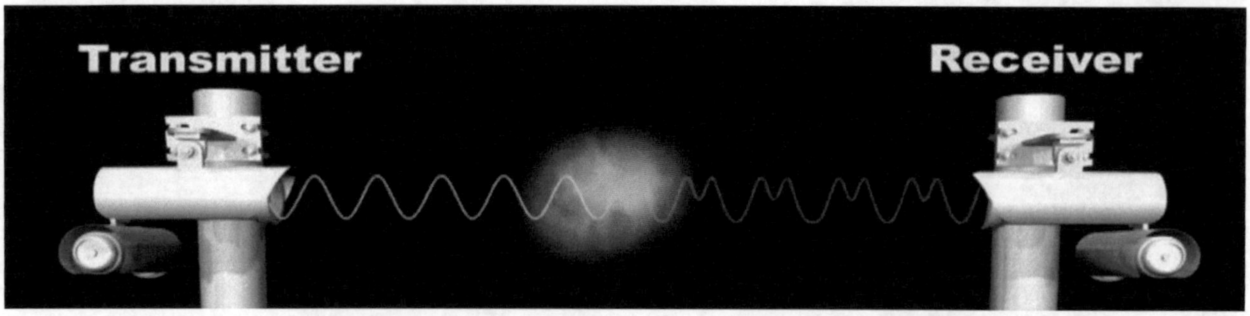

Figure 6. ELDS™ systems use a transmitter-receiver configuration to detect and measure gas concentrations at specific target gas absorption wavelengths over distances of up to 200 meters

Figure 7. ELDS™ systems installed in field operations.

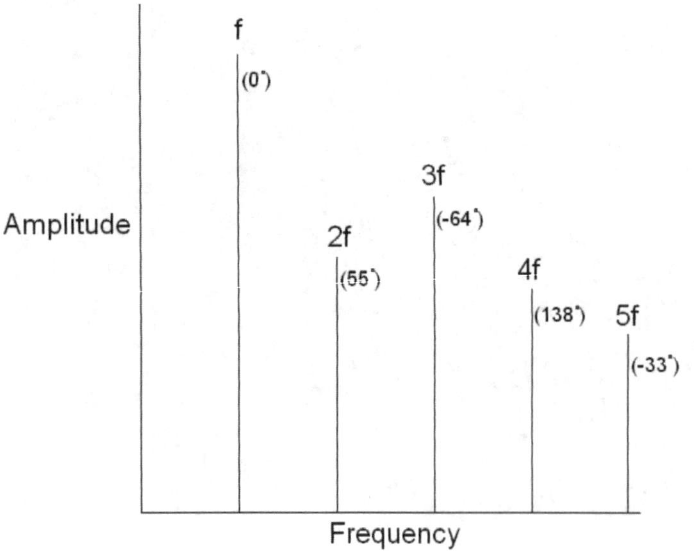

Figure 8. Harmonic Fingerprint produced by scanning a H_2S absorption line at 1589.97nm.

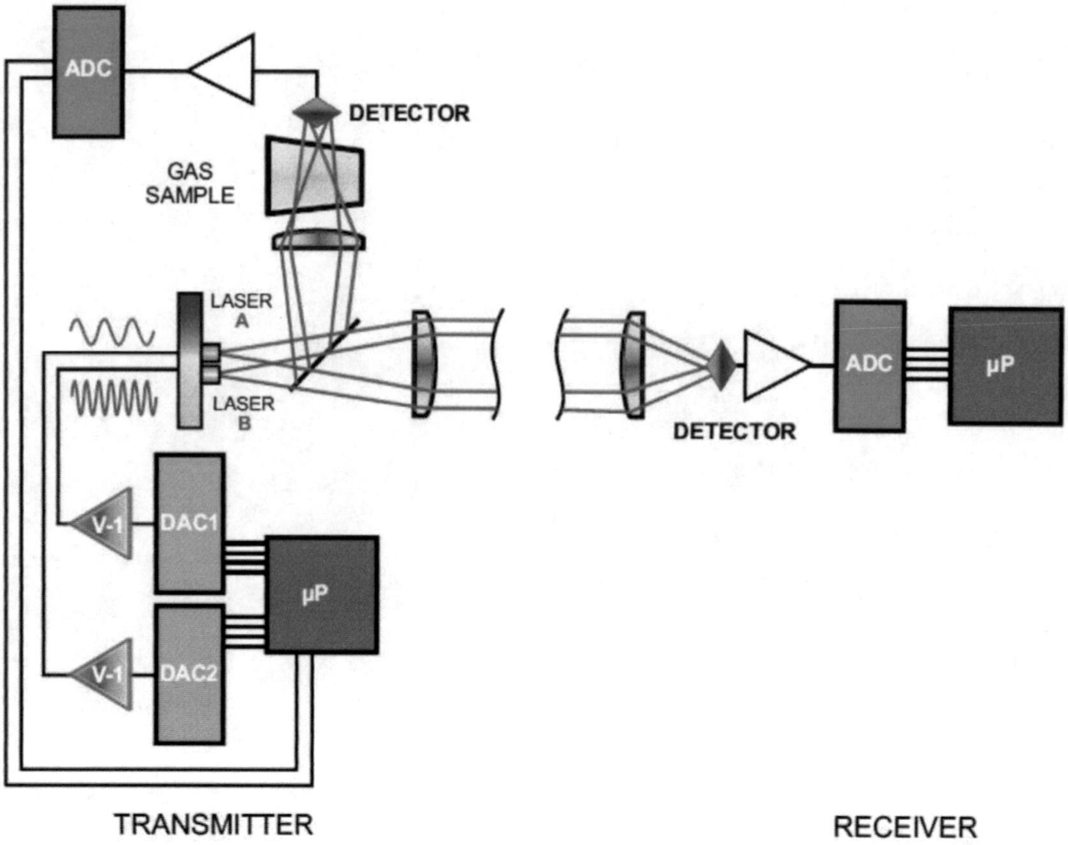

Figure 9. The transmitter unit sends up to two beams of modulated laser radiation through an open path which is to be monitored for the presence of the target gases. The transmitter unit uses its target gas sample to maintain Harmonic Fingerprint lock of its laser diode(s), and in so doing ensures that any absorption of its output beam by target gases in the monitored path will introduce a Harmonic Fingerprint onto the beam. The receiver unit analyses the detected signal, looking for the Harmonic Fingerprint(s) which it knows will be produced by target gases in the monitored path. The receiver unit measures the amplitude of any Harmonic Fingerprint components in the signal reaching it to determine how much, if any, target gas there is in the monitored path and sends a signal to the control system.

Figure 10. Illustration of a gas release simulation resulting in a hazardous flammable cloud scenario (1,250m^3) in the power generation module (M09). The dimensioning cloud sizes for this modules 695m^3. Concentration contours from 50% LEL (lower explosive limit) depicted in deep blue, to UEL (upper explosive limit) depicted in dark red, are shown.

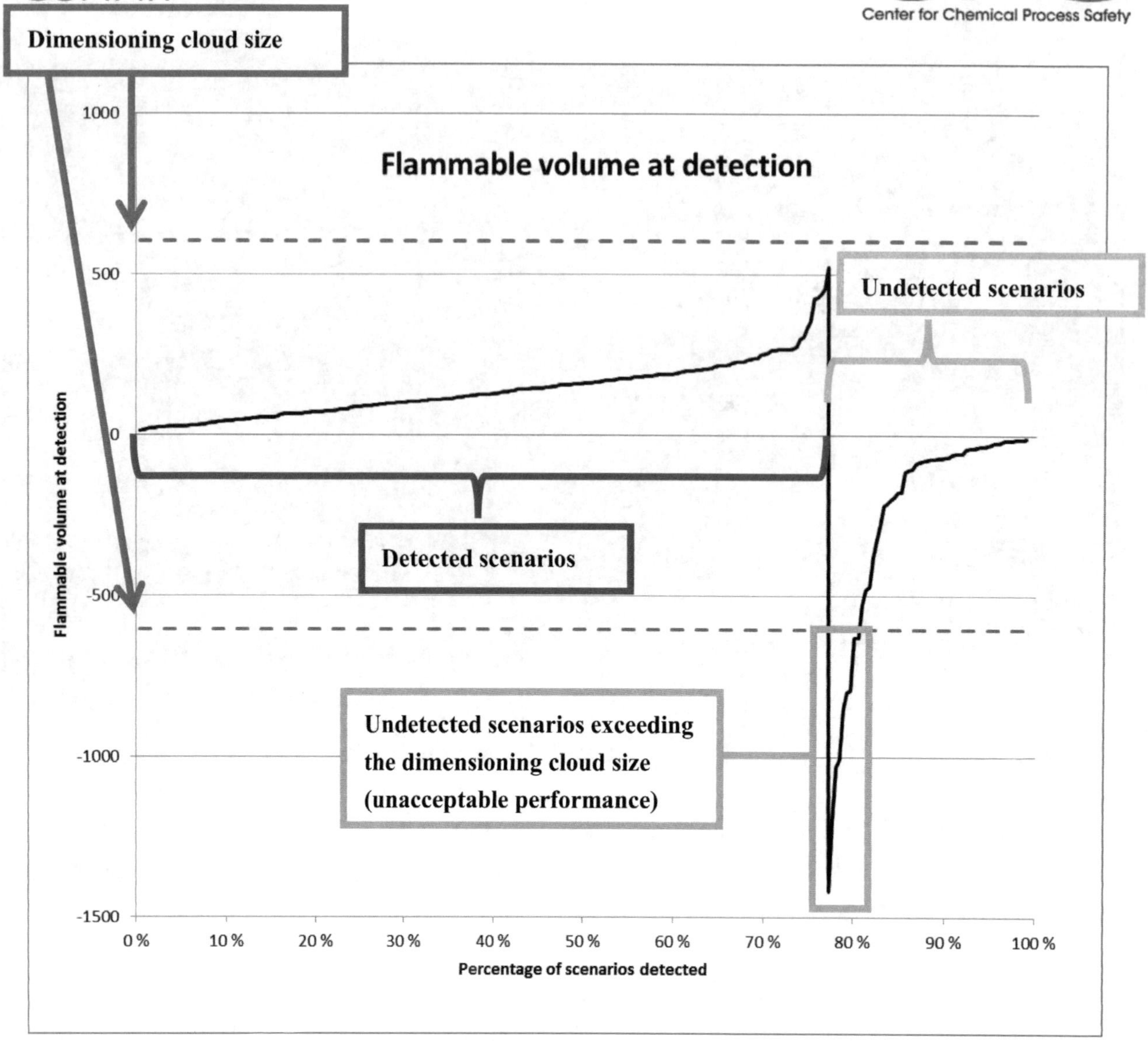

Figure 11. This is a sample graph that depicts important parts of cloud size at detection. A curve which dips below the lower red line represents undetected scenarios over the threshold size, which means the GDS does not satisfy the performance criterion. If the curve crosses the upper red line, this represents scenarios which are detected, but after the cloud reaches threshold size.

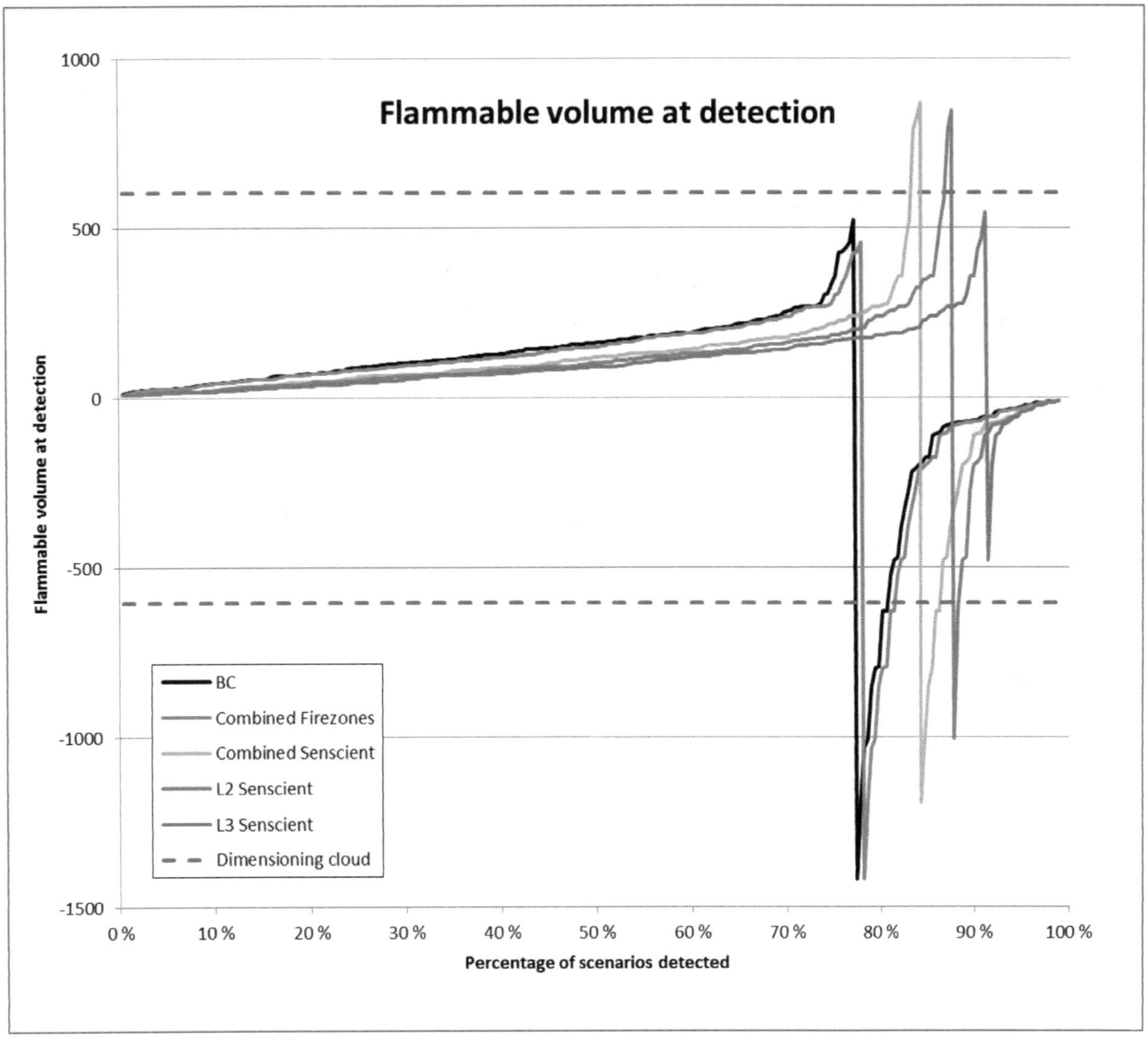

Figure 12. Gas detection simulations and optimization results for module M02 (water injection area). The GDS solution illustrated by the blue curve entitled "L3 Senscient" was selected for implementation. With this solution, all gas leak scenarios bigger than $604m^3$ (the dimensioning cloud size for module M02) are detected before reaching that size. The curve does not cross any of the red lines and satisfies the criterion for hydrocarbon detection.

Figure 13. An ELDS™ transmitter unit installed on the main deck of the Terra Nova FPSO.

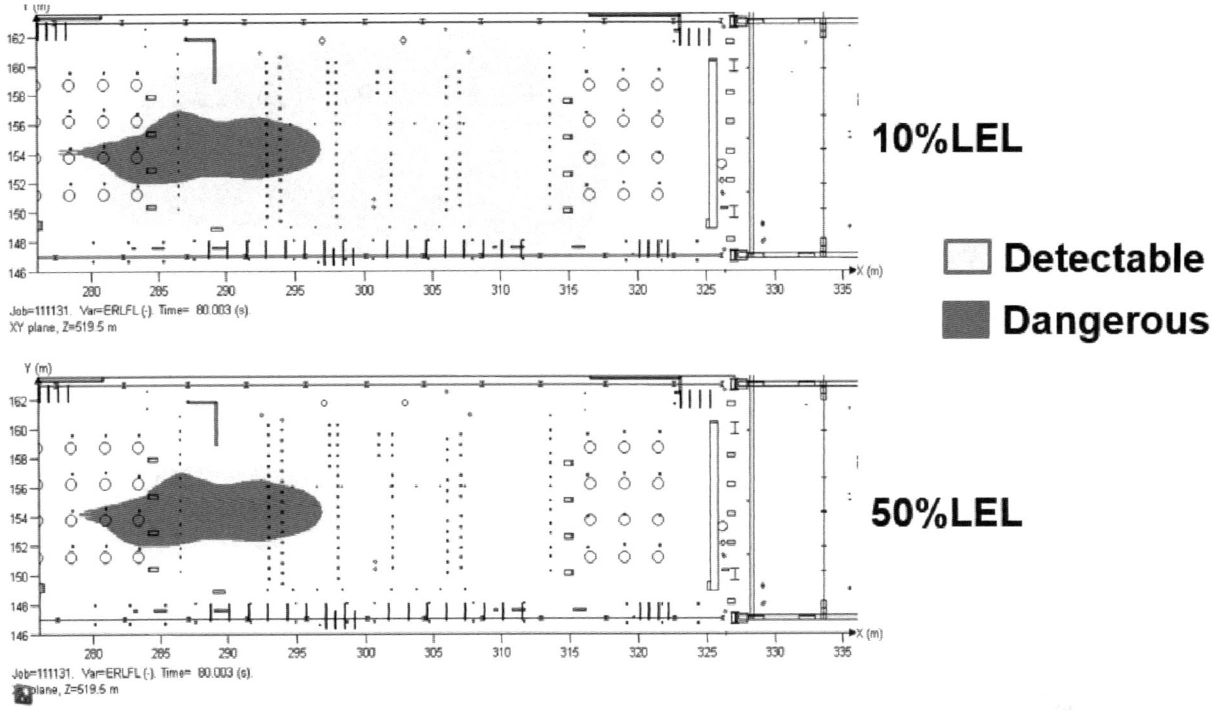

Figure 14. Flammable and detectable volume for a simulated gas leak scenario for two different detection thresholds.

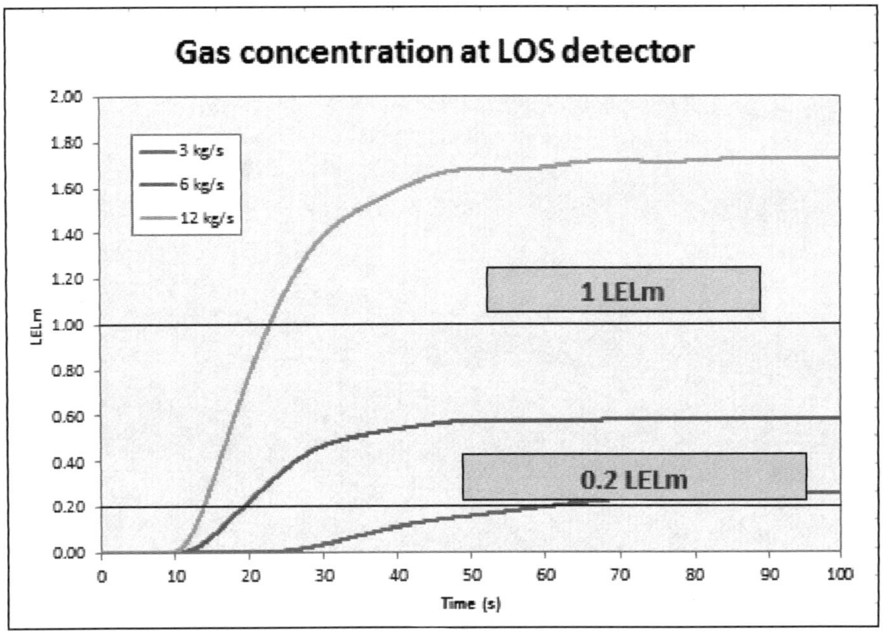

Figure 15. This graph shows the response of an LOS detector to several simulated gas leak scenarios (only leak rate is varied between these scenarios). For the 12 kg/s leak, the difference between a 1 LELm and a 0.2 LELm threshold is a difference in detection time of 5-10 seconds.

For the 6 and 3 kg/s leaks, however, a detector with a 1 LELm threshold will not give alarm even as the cloud reaches steady state.

Enhanced Application
&
Sharing of Lessons learned

RUSKIN DAMANI

RELIANCE INDUSTRIES LIMITED.
HAZIRA MANUFACUTRING DIVISION
SURAT
Ruskin.damani@ril.com

Prepared for Presentation at

UNPUBLISHED

Enhanced Application
&
Sharing of Lessons learned

RUSKIN DAMANI

RELIANCE INDUSTRIES LIMITED.
HAZIRA MANUFACUTRING DIVISION
SURAT
Ruskin.damani@ril.com

Keywords: enter a few keywords here to aid searching

Abstract

While easy access to informative reports of number of incidences including catastrophic incidents poses a cheapest opportunity to learn and improve, it is a challenge to identify appropriate learning looking at organization process and procedures in place and publicize lesson learned. Moreover, there are also challenges due to cultural barriers in accepting lessons and sharing of lessons learned.

Strong management commitment to knowledge sharing, developing a well-defined business plan for implementing knowledge management, providing incentives to encourage knowledge sharing, and building technology systems to facilitate information access and dissemination are critical to the success of knowledge management.

This paper covers RIL's experience on effective utilization of various tools like safety contact, safety observation program, the video films on event, focused week/month celebration, skit, drama, etc. to effectively communicate relevant information to relevant persons of organization such that the appropriate prompt corrective actions are taken and the sense of vulnerability is kept alive. The Governance and Review process around this helps to sustain and improve.

In overall, this paper covers the experience, learning and major challenges in sharing and implementing of lessons learned for improving and sustaining Safety performance at Reliance Industries Limited (RIL) India. Over the years, RIL has taken several strategic steps to make a quantum jump in Health, Safety & Environment (HSE) management systems, practices, performance and culture.

1. Introduction

The Reliance Group, founded by Dhirubhai. H. Ambani (1932-2002), is India's largest private sector enterprise, with businesses in the energy and materials value chain. Group's annual revenues are in excess of US$ 66 billion. The flagship company, Reliance Industries Limited, is a Fortune Global 500 company and is the largest private sector company in India.

Backward vertical integration has been the cornerstone of the evolution and growth of Reliance. Starting with textiles in the late seventies, Reliance pursued a strategy of backward vertical integration - in polyester, fiber intermediates, plastics, petrochemicals, petroleum refining and oil and gas exploration and production - to be fully integrated along the materials and energy value chain.

The Group's activities span exploration and production of oil and gas, petroleum refining and marketing, petrochemicals (polyester, Fiber intermediates, elastomers, plastics and chemicals), textiles, retail, Infotel and special economic zones.

Reliance enjoys global leadership in its businesses, being the largest polyester yarn and fiber producer in the world and among the top five to ten producers in the world in major petrochemical products.

RIL has embarked this journey of process safety excellence to be a leader in this field true to its vision of "Growth is Life" and becoming world-class in every endeavor it takes. Understanding the importance of Process Safety in the businesses it operates, the RIL management has taken several strategic steps to become world class in HSE management. In order to make a rapid quantum jump in HSE management systems, practices, performance and culture RIL has learnt from experiences and knowledge available on process safety through CCPS and has also partnered with one of the global leaders in safety management to transform and become world leader in HSE management.

2. The challenge :

In era of information technology and active media, information spread about industrial incident which has taken place anywhere in world is in fraction of second. Incidents taking place around chemical industry is continuously posing sense of vulnerability and unease. Though catastrophic incidents are rare; the impact of such incident is hue. Looking at such catastrophic incidents, all stakeholders raises questions to industries.

Some of Institute like Chemical safety board (CSB) carries out the detail incident investigation and publish report and animation videos. While easy accessibility to such informative detailed report helps one to evaluate and identify opportunity for improvement, it also put forward a challenge to identify the applicable learning for the organization.

Like any organization, we at RIL also continuously look forward to adopt learning from the industrial incidences across globe to prevent any such incident as learning from incident from other industry is the cheapest way to learn and improve the system.

With implementation of safety management system at RIL, an IT enabled incident and near miss reporting system is established across RIL sites. Any employee including contractor can report the near miss/Incident in system. The incidents are investigated and investigations reports of near misses/ incidents are made easily accessible through this IT enabled system.

The incident investigation report often indicates that there is nothing new. Incidents across world indicate that well defined management systems and their associated policies and procedures ultimately depend on the action of individuals and groups for their successful implementation. The learning from incident is not effective enough to prevent similar incidents. Thus, the major challenge lies in disseminating relevant learning from incidents to relevant employee and adopting learning as appropriate in preventing incident of similar nature.

Employees of different background, different culture are working at site. Significant number of work force is contractor employees whose turnover is high. This also affects on sustaining sound safety culture.

3. Adoption Of Different Approaches

Actions of the respective individual working within the system are responsible for ensuring and sustaining sound safety culture. These cultures have been developed in response to, and are reinforced by frequent reference to, significant loss events across the world. It is beneficial to draw upon the experience of others in there, or similar industries. At RIL, So as to enhance effective learning from incident and promoting learning culture to develop and sustain sound safety culture, strategic approaches were deployed.

a. Preparation of Learning for incident investigation and circulation

After preparation of near miss/ incident investigation report, learning from incident (LFI) is prepared by the investigating team leader in the iRIL system. Once released in system, through automatic mail, it is accessible across all sites. The site representative reviews and identifies learning for appropriate corrective actions. While reviewing the LFI following steps as mentioned in fig.1. The action emerged are assigned to concerned through action tracking system and same is being monitored till closure through system.

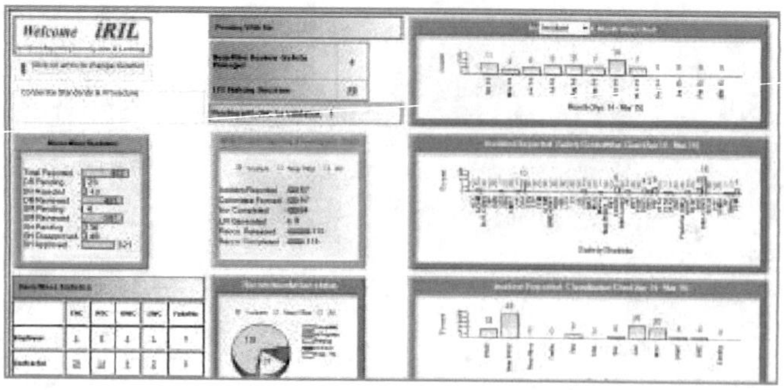

The Learning from incident is also shared with all concern at site through various forums like E-mail, Display at designated location, discussed in meetings as a safety contact.

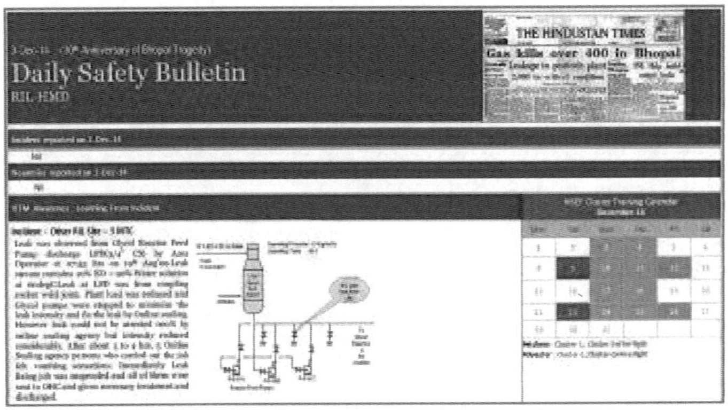

When the incident is of higher consequences or incident of other industry the learning of incident is prepared by Center of health, safety and environment excellence (CHSEE) and is being circulated to all sites. Such LFIs are also discussed, debated at different forums for identifying relevant learning & actions emerged are worked upon. In this process also, steps as mentioned fig.1 are followed. IT enabled system helps to communicate the learning from incidents at speed.

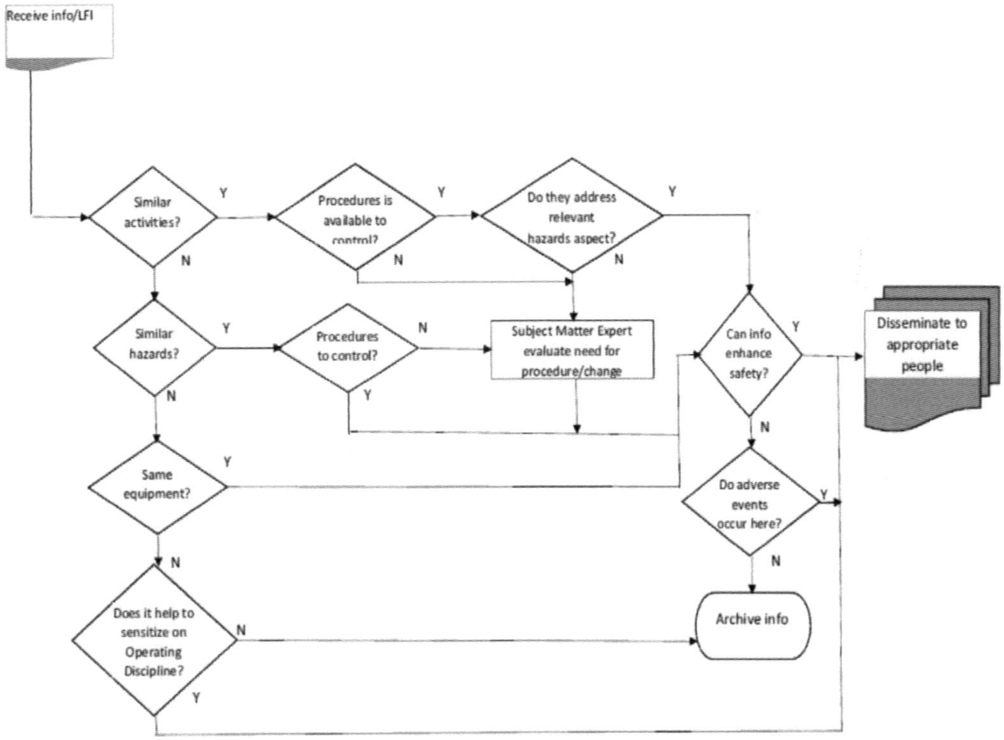

This methodology has helped to identify applicable physical and systemic change at the site and implement as required for preventing similar incidents and educate importance of each change with relevant correlation. However, at occasion, influence on human behavior was not found sustainable as humans tend to forget learning over period.

b. Safety observation program

At Reliance, seven steps safety observation is designed to improve safety behavior of work force to build sustainable safety culture. During safety observation round leader interacts with work force in one-one basis or with small group of people. It gives opportunity to share the appropriate learning from past/recent incidents having relevance to their domain. As the conversation takes place in regional language and in cordial informal manner, there is openness. This helps listening feedback /issues from workforce as well as spreading relevant learning from incident as well. At the end of safety observation process, an agreement is made with respect to implementation of action item if any and commitment towards compliance to lay down safe working practice/procedure. This process has helped to improve safety culture as well as disseminating appropriate learning from incident.

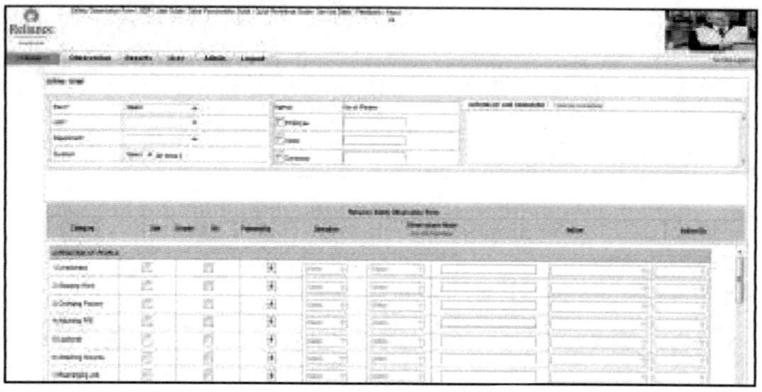

This methodology has helped to disseminate the relevant learning from incident correlating the existing specific issue in vernacular language with front line employees/Shop floor employees. Normally in this process, two way communications takes place which helps to ensure understanding of recipient. In this method, Effectiveness of this varies as it totally depends on individual employee carrying out safety observation round.

c. Screening of safety videos of same/ different level of employees of organization

Seeing is believing. Animation videos available for different incidents depict the sequence of events and the lapses in the systems leading to event. Such videos are viewed with targeted audience of different combination i.e. employees of same plant, employees of same discipline, employees of different discipline, employees of different levels.

After screening of such safety videos, an interaction session is held covering various aspect like "what went wrong", "how it cannot happen here", "what are our systems", "what are the suggestion to prevent similar type of incident at our plant/site". This helps to sensitize and keep the sense of vulnerability amongst group. This also helps to identify opportunity for improvement as well as emphasize on importance of present safety system in place.

Using this method, we achieved comparatively better results as it helps to brainstorm issues amongst employees of all disciplines. This has helped to improve importance of ensuring compliance to prevailing standard/system ongoing basis and identify opportunity for improvement and work in time bound manner.

d. Skit / Drama

Skit is a one of strong communication tool to spread relevant message to masses in a short time. Taking theme from learning from incident, employees from

workforce team prepare a skit and conduct program at learning center of site or common place where many people together can see. As it is a humorous way of conveying message, this proves one of the best ways to communicate relevant learning from incident front line employees. Using this method, we could increase employee engagement in spreading the learning and improve implementation as well as individual behavior.

e. Celebration of safety week/Month

Taking a specific theme for the program, a safety week or safety month is being celebrated every year. During this period, banners containing relevant messages are displayed at strategic locations. Various competitions like safety talk, competition is being conducted to spread message across all level of employees. Every time some unique features are added to sustain employee engagement and avoid monotony.

4. Conclusion

It is always a challenge to identify learning applicable to site / plant /group's most pressing issues in a positive manner and inspires people to take prompt action from the information available. Sharing relevant aspect using appropriate tool to appropriate audience can expand the breadth and depth of the system as well as provide emphasis on importance to comply with existing laid down system.

We at Reliance had used various applications to enhance sharing of lesson learned through its safety excellence journey. This helped us to share learning, update knowledge and/or re-emphasize the "desired value" as well as "keep sense of vulnerability of alive" without losing confidence on our effort.

There were many successes and some failures. With every success the goalpost of safety excellence had moved further as it is a continuous Journey. The failures made us to learn and enrich us by strengthening the system to avoid future mistakes.

Whatever one does it, it is important to keep the organization, at all levels, focused on "what is at stake". At particular instant one method proves superior whereas other method proves better at next instance. Often combination of method proves better. There is no single method is best. Keeping this in mind we continuously strive for evolving different approaches different time remaining focus on sharing of learning.

Case Study:-I

CSB safety video "Anatomy of Disaster" screening sessions were conducted with different discipline representative, same dept.employees, different level of employees etc. It was followed by discussion on

- ✓ How it is relevant to us?
- ✓ How it can be avoided?
- ✓ Importance of PSM.
- ✓ Importance of individual role and responsibility…. etc.

Based on above detailed actions plan was prepared and worked upon and number of improvements were made in existing system/practices.

Sample appended below.

Summary of BP Texas findings and the actions identified based on review in respect to existing system/ practices

➢ Technical finding

Sr	Major Finding	Major learning	Existing System /Practice	Action Identified.
1.	The tower level indicator showed that the tower level was declining when it was actually overfilling. The separate high level alarm did not activate, and the tower was not equipped with any other level indications or automatic safety devices.	Effectiveness of PM system of instruments	1. Instruments are mapped in SAP-PM schedule and being followed.	-
2.	A lack of supervisory oversight and technically trained personnel during the startup, especially hazardous period, was an omission contrary to BP safety guidelines. An extra board operator was not assigned to assist, despite a staffing assessment that recommended an additional board operator for all ISOM startups.	Capable supervision during critical operation like start-up and shut-down.	1. Capable and sufficient resources are being ensured as part of planning and execution during start-up and shut-down. 2. Shift-in-charge is capable to take safe shut-downs in case of emergency. 3. Experts / leaders are available 24x7, on	1. Review of critical shift positions adequacy to handle plant emergency situation, in view of additional facilities / DBN.

			call if required during silent hours. 4. In case of requirement, shift-in-charge is authorized to hold back shift personnel for additional support.	
3.	Supervisors and operators poorly communicated critical information regarding the startup during the shift turnover; BP did not have a shift turnover communication requirement for its operations staff.	Proper communication and HOTO during shift-change	1. Formal HOTO protocol is being practices at each level – e.g. shift-in-charge, Operator etc. 2. In case of personnel emergency of any individual, competent substitute being arranged. No person can leave duty post without replacement.	1. Study available time for communication and increase time for communication during HOTO if required. 2. Study available time for communication during HOTO while plant is under decommissioning for shutdown / start up after major shutdown.
4.	ISOM operators were likely fatigued from working 12-hour shifts for 29 or more consecutive days.	Continuous work without off.	1. This is being taken care during shift schedule preparation and execution stage.	-

➢ Organizational Findings:

Sr.	Major Finding	Relevance	System / practice at HMD
1	Cost-cutting, failure to invest and production pressures from BP Group executive managers impaired process safety performance at Texas City	Importance of process safety management.	• Extensive efforts are deployed for nurturing PSM culture.
2	The BP Board of Directors did not provide effective oversight of BP's safety culture and major accident prevention programs. The Board did not have a member responsible for assessing and verifying the performance of BP's major accident hazard prevention programs.	Management commitment.	• Site management / HSE subcommittees are closely monitoring safety systems / culture.

3	Complacency due to low personal injury rate at Texas City as a safety indicator failed to provide a true picture of process safety performance and the health of the safety culture.	Monitoring process safety performance data along with injury statics is essential.	• Being monitored at RIL. • PFDE incidences are being tracked. • LFI being shared / implemented.

By this exercise, following system were further improved.

- Hand-Over-Take-Over systems at shift change over especially during Shutdown.
- Plant Startup SOP's and SOCs limits based on past startups
- Strengthening system of refreshing shutdown / start up SOPs before major annual turn around
- Safety valve sizing calculation revalidation for cases where proprietary information is involved.
- Porta-Cabins by carrying out facility siting.

Actions are in hand.

- Relocation of fire station/OHC and Main administration building based on facility siting study.

Case Study-II

Inert Entry-

Based on learning from Inert Entry failure of Other Industries, we have implemented Practice of using Lockable helmet instead of Respiratory Protection Mask. To strengthen further, Life Support Panel provided is kept separate for both Entrant and Rescuer. Although cost of Contract due to these changes was almost double than earlier practice, still these changes are implemented in all Inert Entry jobs.

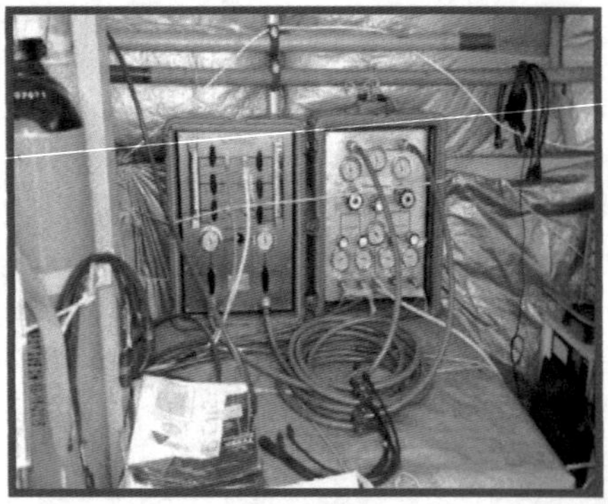

Separate Life Support Panel for Entrant and Rescuer

Lockable Hood Type Helmet used in Inert Entry Jobs

NTNU – Trondheim
Norwegian University of
Science and Technology

A new assessment method for regularity and safety in oil and gas production

Yiliu Liu
Department of Production and Quality Engineering

13.01.2015

www.ntnu.edu

2

A brief introduction

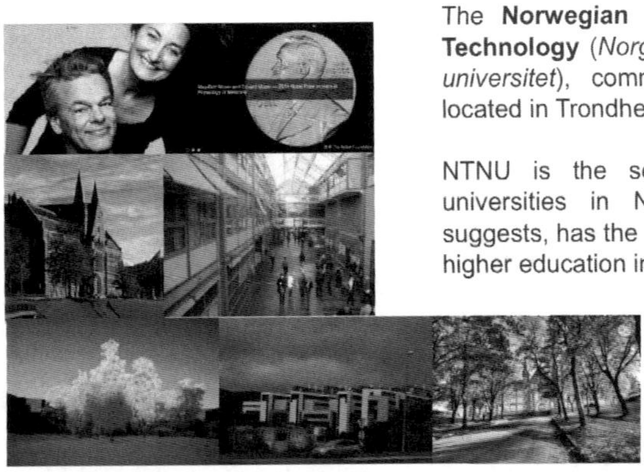

The **Norwegian University of Science and Technology** (*Norges teknisknaturvitenskapelige universitet*), commonly known as **NTNU**, is located in Trondheim.

NTNU is the second largest of the eight universities in Norway, and, as its name suggests, has the main national responsibility for higher education in engineering and technology.

NTNU – Trondheim
Norwegian University of
Science and Technology

www.ntnu.edu

3

A brief introduction

NTNU – Trondheim
Norwegian University of Science and Technology

www.ntnu.edu

4

RAMS group

RAMS (Reliability, Availability, Maintainability and Safety) is in the department of production of quality engineering.

Academic staff: 4 professors, 2 associate professors and 2 adjunct professors

Our aims: To be recognized as an internationally leading research group on selected topics within the RAMS area, and to offer world-class education in RAMS.

Probably the largest RAMS group at any university.

NTNU – Trondheim
Norwegian University of Science and Technology

www.ntnu.edu

5

Prioritized research areas

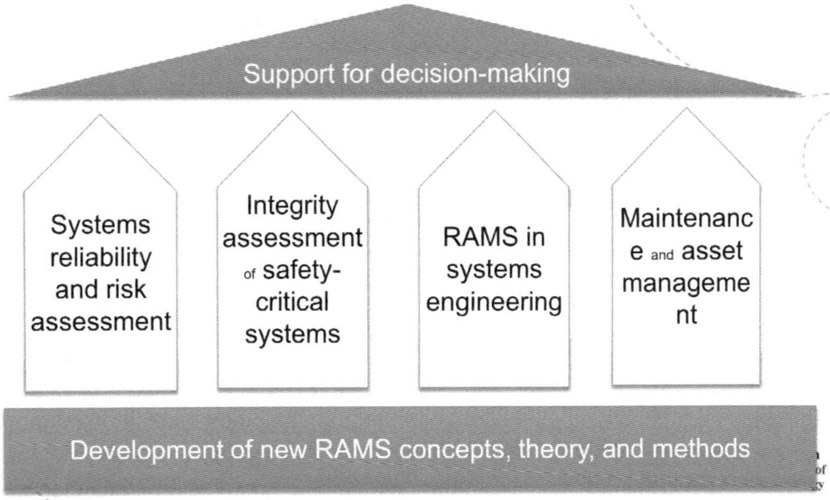

www.ntnu.edu

6

Books

New:
Reliability Engineering…

Y. Liu & M. Rausand, 2015

NTNU – Trondheim
Norwegian University of Science and Technology

www.ntnu.edu

7

Education – master programs

1. Master in Product Development and Production (Mechanical Engineering) – 5 years program

2. International Master in RAMS – 2 years program (by English)

3. Master in Subsea Technology – 2 years program

NTNU – Trondheim
Norwegian University of Science and Technology

www.ntnu.edu

NTNU   Studies   Living in Trondheim   Research   About NTNU   Contact us

Search

› Reliability, Availability, Maintainability and Safety (RAMS)

- About the programme
- Programme structure
- Job prospects
- Study environment
- Student guidance
- Admission
- Why study RAMS
- Learning outcome
- Courses - Quick Reference

MASTER'S DEGREE PROGRAMME, 2 YEARS

MSc in Reliability, Availability, Maintainability and Safety

— About the programme

Can we accept that a small flaw can cause a car's brakes to fail? Can we deter those dangers to both people and the environment that are inherent in the complex products and systems surrounding us?

The objective of this Master's programme is to provide knowledge and skills to enable students in the development and operation of safe, reliable and easily maintainable systems. You will learn how products and systems can be used safely, and how technical faults can be avoided. Furthermore, you will learn how to plan and cost-effectively perform maintenance. The language of instruction **is English.**

This programme is unique in Scandinavia and builds on long experience from industry projects, teaching and research. The study is suited for people who like to identify and solve problems, have analytical skills, can work structured and systematically, and enjoy both theoretical and practical problems.

The study environment

NTNU and SINTEF together constitute the largest academic environment in Europe within safety, reliability and maintenance, and has established the ROSS Gemini Centre. Read more about the study environment.

Why study RAMS?

Product and system complexity is increasing. Many technical products and systems are combinations of software, electronics and mechanical components. Also, the interactions between technical systems and the economical / social systems of our global society are becoming tighter and more inter-dependent. More on Why study RAMS?.

Personnel Transport © Statoil

Flytoget Oslo © Flytoget AS

Statfjord C © Statoil

Brochure about RAMS

Download a brochure (pdf) about the RAMS-group, research activities, study programs, and courses.

FACTS:

Level / Degree type:
International Master's (MSc) - 2 yr.

Language of instruction:
English

Programme code:
MSRAMS

Number of students admitted:
15

Host Faculty:
Engineering Science and Technology

Host department
Production and Quality Engineering

Application Deadlines

15 JANUARY: International applicants
The online application is now open for self-financed and Quota Scheme applicants.

1 MARCH: Nordic applicants
The 1 March deadline may apply if you have a bachelor's degree from a Nordic country or are a Nordic citizen. See admission procedures.

15 APRIL: Norwegian applicants
If you are a Norwegian citizen and hold a bachelor's degree from Norway. See admission procedures.

http://www.ntnu.edu/studies/msrams

9

RAMS continuing education

Course	Spring	Fall
PK6017 Risk analysis	X	
PK6018 Safety and reliability analysis	X	X
PK6019 Maintenance management	X	
PK6021 Maintenance optimization		X
PK6022 Reliability of safety critical functions		X
PK6026 Technical Safety	X	

NTNU – Trondheim
Norwegian University of
Science and Technology

www.ntnu.edu

10

A new assessment method for regularity and safety in oil and gas production

NTNU – Trondheim
Norwegian University of
Science and Technology

www.ntnu.edu

11

Background

- Production regularity is a concept used in Norwegian oil and gas industry, as the actual production divided by the maximum possible production (in the relevant period of time)
- The probability distribution of throughput capacity (PDC) is used as the quantitative measure of the production regularity
- PDC is not a physical measure, but means the percentage of a specified throughput

NTNU – Trondheim
Norwegian University of
Science and Technology

www.ntnu.edu

12

Objective

- Safety assessment is also very important, as well as reducing the frequency of accidents and mitigating the accident consequences
- The objective of this paper is to include the safety integrity assessment in the measurement of PDC. Reliability block diagram driven Petri net will be used in modeling of the operation of components
- The overall system is regarded as a structure with three levels: components, sub-systems and system

NTNU – Trondheim
Norwegian University of
Science and Technology

www.ntnu.edu

13

Objective

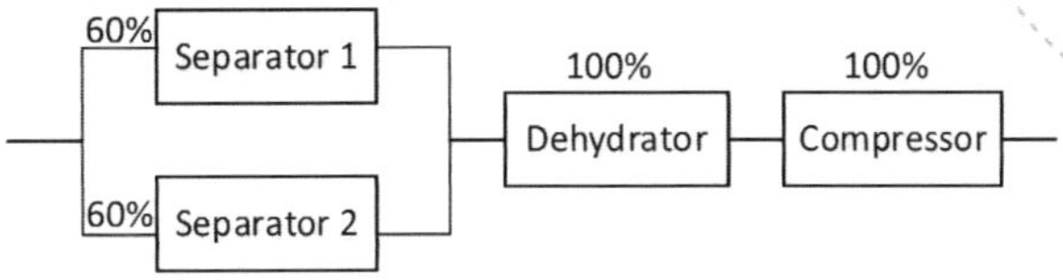

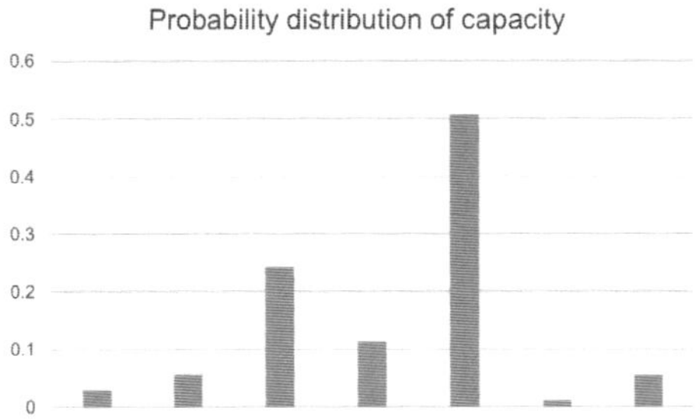

14

Merging rules

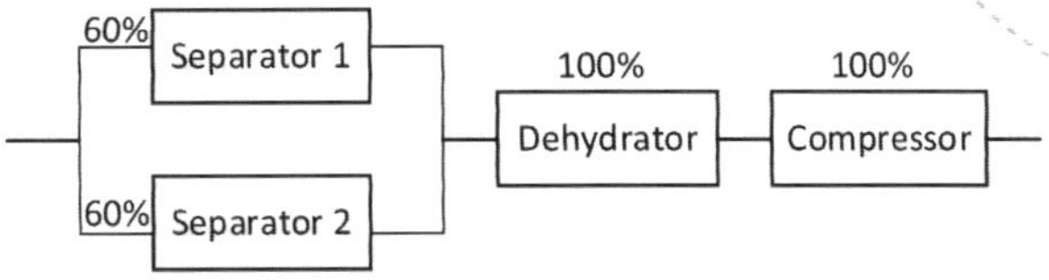

For a series structure

$$\Pr(Z = z) = \Pr(X = z)\Pr(Y \geq z) + \Pr(Y = z)\Pr(X > z)$$

For a parallel structure

$$\Pr(Z = z) = \sum_{x,y,x+y=z} (\Pr(X = x)\Pr(Y = y))$$

15

Component modeling

- A Petri net model for a component with four throughput levels, e.g. 100%, 70%, 40%, 0%.
- The associated SIS has two states: functional and failed

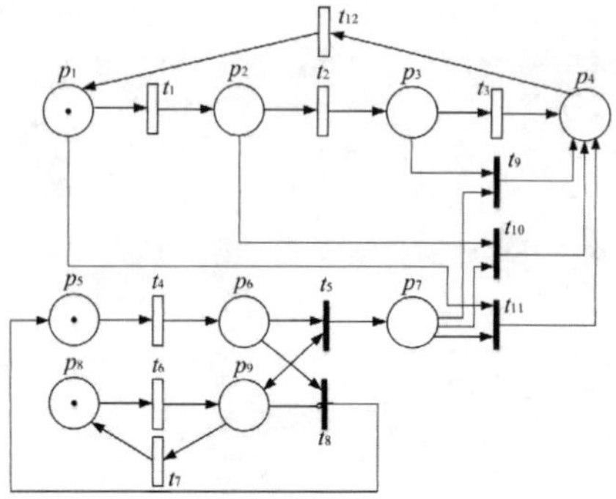

16

Case study

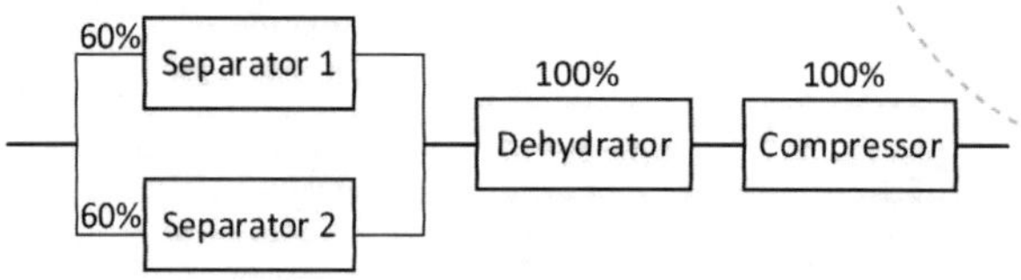

- The dehydrator and compressor have four throughput capacities: 100%, 70%, 40% and 0%
- Separator has three levels: 60%, 30% and 0%.
- *The possible throughputs of the system can be found as 100%, 90%, 70%, 60%, 40%, 30% and 0%.*

- The degradation rates are 5×10^{-4} per hour.
- For SISs, their failure rate are supposed as 5×10^{-6} per hour
- Proof tests are carried out once per year.
- Demand rate is 10^{-5} per hour.
- The restoration rate of production equipment is 0.01 per hour.

17

Case study

The PDC of the overall system

Throughput capacity level (%)	PDC
100%	0.0294
90%	0.0562
70%	0.2414
60%	0.1133
40%	0.5059
30%	0.0111
0%	0.0541

- Petri net models can be solved with simulation method in many software, and here we select GRIF. The iteration time is 5 years.
- The probability of accidents of the overall system is 7.7196× 10^{-7}.

NTNU - Trondheim
Norwegian University of Science and Technology

www.ntnu.edu

18

Summary

- This paper employs Petri net to capture the probability of throughput capacity distribution and the probability of accidents of equipment, and then obtain the PDC and PA of the overall production system based on the specific merging rules.
- The future studies can be carried out on the continuous throughput distribution of equipment. In addition, the common cause failures in the SISs and equipment also should be taken into consideration.

NTNU - Trondheim
Norwegian University of Science and Technology

www.ntnu.edu

19

Thanks a lot for your attention

More questions: http://www.ntnu.edu/ross/rams

NTNU – Trondheim
Norwegian University of
Science and Technology

www.ntnu.edu

EVALUATION OF INDUSTRIAL RISK ASSESSMENT FRAMEWORK – AN INDIAN PERSPECTIVE

2014 CCPS Global Summit on Process Safety

15-16 December 2014, Mumbai, India

Anandita Sengupta

PhD Student - University of Twente, The Netherlands

Safety Analyst, Safety Centre, Indian Institute of Technology Gandhinagar

UNIVERSITY OF TWENTE.

Outline

- Background
- Introducing Seveso Directive
- Risk assessment from Indian perspective

UNIVERSITY OF TWENTE.

Why India?

- The worst industrial accident occurred in Bhopal in December, 1984
- Exposure to MIC, leaking from Union Carbide Plant, cause significant number of deaths and injuries
- Caused a world wide rethink on risk from hazardous installations

UNIVERSITY OF TWENTE.

What resulted in such high impacts?

- No information made available from Union Carbide on the nature of hazards within premises
- Public was unaware on how to react in order to minimize exposure to Methyl Isocyanate (MIC)
- Hospitals & medical professionals did not know what gas was released and how to treat victims
- There was no emergency response plan which could assist administration to minimize impacts

UNIVERSITY OF TWENTE.

Other examples

- Flixborough, UK 1974
- Seveso, Italy 1976
- San Juan City, Mexico 1984
- Basel, Switzerland, 1986
- Enschede, The Netherlands 2000
- Toulouse, France 2001
- ....
- ....

UNIVERSITY OF TWENTE.

Evolution of the Seveso Directive (I)

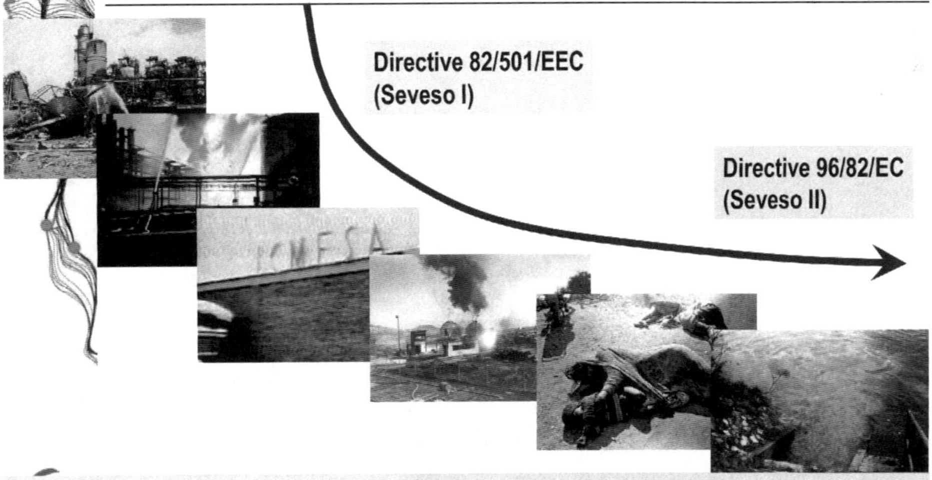

Flixborough (UK) June 1974	Beek (NL) November 1975	Seveso (I) July 1976	Mexico City (MEX) November 1984	Bhopal (India) December 1984	Basel (CH) November 1986

Evolution of the Seveso Directive (II)

Amended Directive "Seveso II" (2003/105/EC)

Reviewing of the current Directive

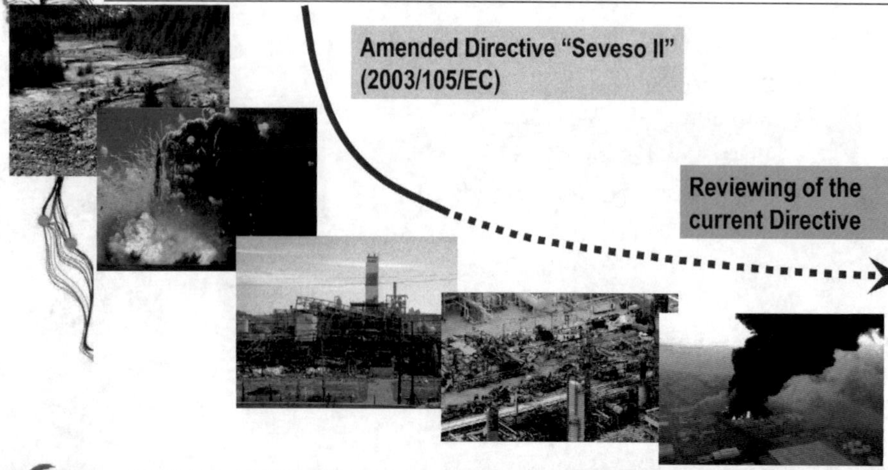

Baia Mare (RO) January 2000	Enschede (NL) May 2000	Toulouse (F) September 2001	Texas City (USA) March 2005	Buncefield (UK) December 2005

Sprit of the Seveso Directive (I)

Why does the European Union needs a Seveso Directive?

To ensure throughout the European Union a common high level of safety and protection against chemical accidents for the people and the environment.

Is legally required by Article 175 EC of the Treaty establishing the European Community (defines objective Environmental Policy of the Community).

ITC UNIVERSITY OF TWENTE.

Sprit of the Seveso Directive (II)

How can Member States reach the Seveso Directive aim?

By preventing major accidents that involve dangerous substances;

By mitigating the consequences of accidents on man and environment.

What do Member States have to do?

Implementing the Seveso Directive into their legislation because it provides them with a legal and technical framework to fulfil the obligations to the European Union.

UNIVERSITY OF TWENTE.

What has happened to India so far?

UNIVERSITY OF TWENTE.

Post Bhopal Era

- Environmental Protection Act,1986
- Manufacture, Storage, Import of Hazardous Chemicals Rules, 1989
- Emergency Planning, Preparedness and Response to Chemical Accidents Rules, 1996;
- Zoning Atlas for sitting an industry
- Environmental Mapping approach
- National Guidelines on Chemical Disaster Management 2007 (under DM Act 2005)

UNIVERSITY OF TWENTE.

The Concerns

- No integrated system for assessing and managing risk arising out of hazardous industries in India
    - No standard criteria for risk assessment which can be translated into Land Use Planning decisions;
    - Lack of risk acceptability or tolerability criteria
    - Method of accident probability calculation
    - Risk reduction strategy for the areas having high risk
- Decision makers often do not have access to risk information.

UNIVERSITY OF TWENTE.

Why a comprehensive risk assessment framework is important for India?

UNIVERSITY OF TWENTE.

The concerns

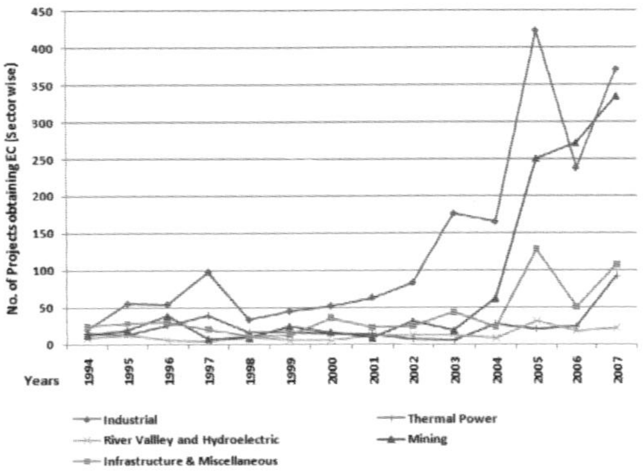

- High rate of development in last few decades
- Doubling in GDP in 7 years - significant reduction in poverty
- Economic growth is sustained in spite of global recession

UNIVERSITY OF TWENTE.

The concerns

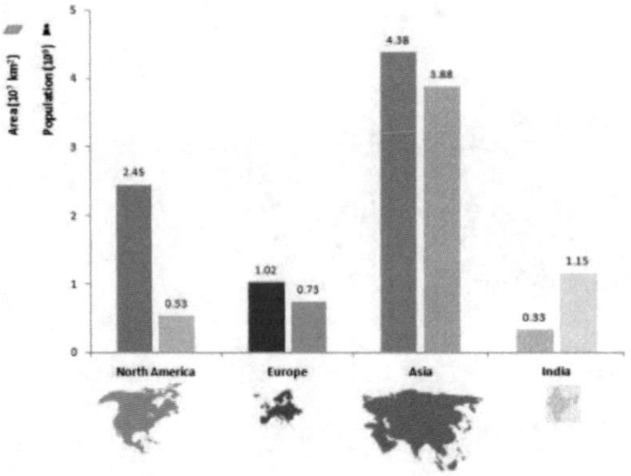

Source: United Nations World Prospects Report, 2004

- High population densities in Asia & India
- Area availability per person is low

UNIVERSITY OF TWENTE.

The concerns

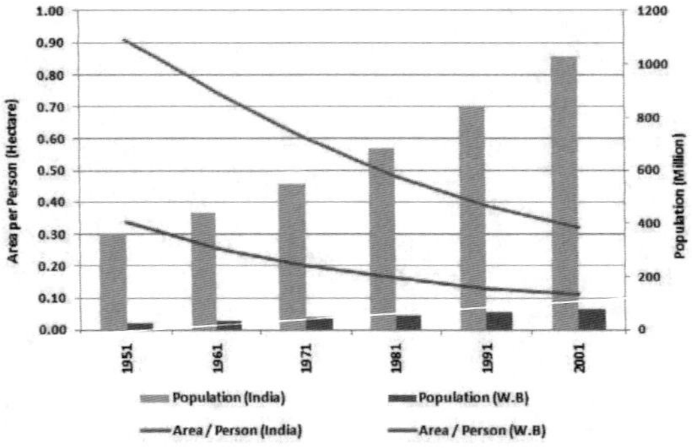

Source: Dept. of Municipal Affairs, Govt. of WB

- High population densities in Asia & India
- Area availability per person is low
- Steady growth of population

UNIVERSITY OF TWENTE.

The concerns

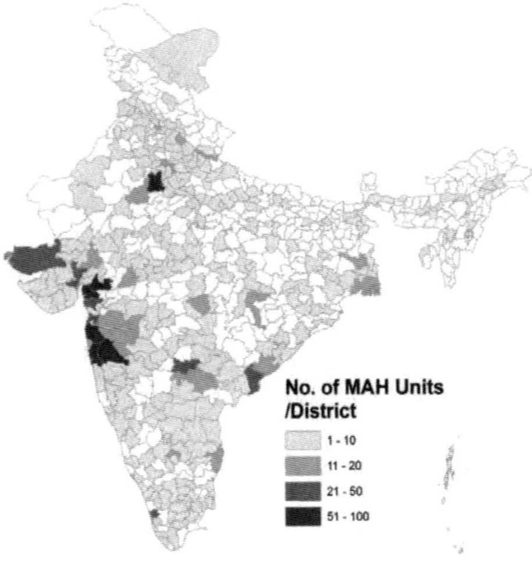

- More than 150 industrial clusters in India have concentration of MAH industries;
- Industries in most of these clusters have come up in an unplanned manner over time;
- Such areas have also witnessed high growth of population over time.

UNIVERSITY OF TWENTE.

Appraisal of risk assessment system of India

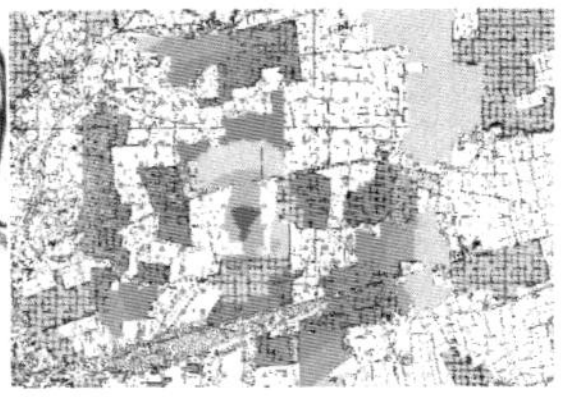

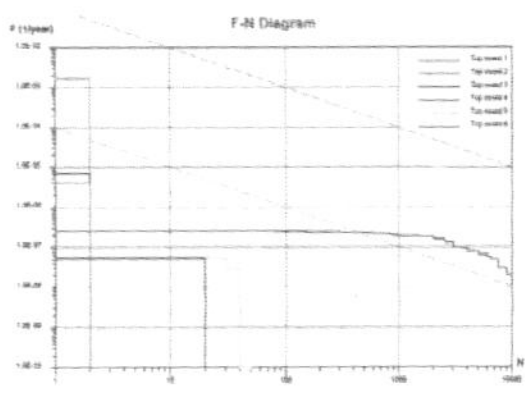

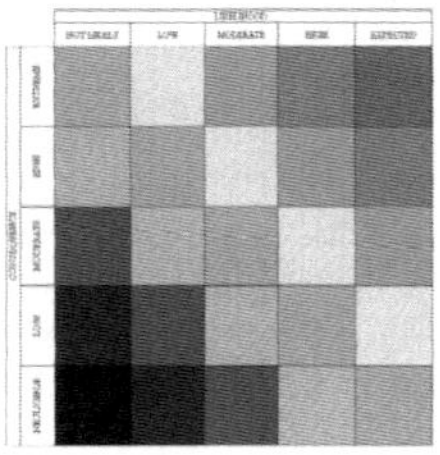

UNIVERSITY OF TWENTE.

The context

- Policy Framework
- Regulatory framework
- Institutional Framework
- Procedural framework
    - ✓ Hazard identification
    - ✓ Frequency estimation - accident database
    - ✓ Arriving at acceptable scenario
    - ✓ Risk analysis
    - ✓ Risk evaluation

UNIVERSITY OF TWENTE.

Policy Framework

- Policy Framework of the Government of India has focused on policy formulation from the sectoral perspective
- Policies in the front of environmental protection and resource consumption
    - ✓ National Conservation Strategy and Policy Statement on Environment and Development
    - ✓ Policy Statement for Abatement of Pollution

The scope was considered to be too broad and not providing any specific direction on how such goals were to be attained. Moreover, in spite of Bhopal disaster have occurred, the policy did not mention any direction that to be adopted for attaining effective management of industrial risks.

UNIVERSITY OF TWENTE.

Policy Framework

- National Environmental Policy (NEP) 2006
  - ✓ *Lack specific focus or direction on several aspects*
  - ✓ *Many of the principles stated in the policy essentially reads like a listing of global best-intentions rather than being directed at specific focuses which are compatible with existing impact level*
  - ✓ *Specific emphasis was provided to the adoption of the 'precautionary principle'*
  - ✓ *However, jurisdiction appears to be more from the point of environmental protection rather than the aspect of industrial risk*
- Disaster Management Policy (DMP) 2009
  - ✓ *Yet again no added focus or strategy for managing the risks that may arise in MAH industrial clusters*

ITC UNIVERSITY OF TWENTE.

Regulatory Framework

Pre-Bhopal

- The Explosives Act, 1884; the Petroleum Act, 1934; the Factories Act, 1948, the Inflammable Substances Act, 1952; the Insecticide Act, 1968 etc.

Post-Bhopal

- Environmental Protection Act,1986
- Manufacture, Storage, Import of Hazardous Chemicals (MSIHC) Rules in 1989
- Emergency Planning, Preparedness and Response to Chemical Accidents (EPPRCA) Rules in 1996 amended 2000
- Besides, the Environment (Protection) Rules 1986 (amended 2004); the Hazardous Waste (Management & Handling) Rules 1989, (amended in 2000 and 2003); the EIA Notification, 2006; etc.

ITC UNIVERSITY OF TWENTE.

Changes in Regulatory Framework

Legislation	Implications
Factories Act, 1948 amendment in 1987	This act includes 11 chapters describing the preliminary information about factory and the duties of the operators.
Environment Protection (EP) Act, 1986	Enacted as umbrella act by central government in 1986 to cover the identified gaps in existing legislation. Rules were subsequently framed under various heads.
MSIHC Rules, 1989 amended in 1994 and 2000	These rules describe the definitions of 'hazardous chemicals', 'industrial activity', 'isolated storage', 'major accident', 'site' etc. In particular deal with the following aspects: 1. Notification of a major accident (Rule 5) 2. Site notification (Rule 7) 3. Updating of site notification (Rule 8) 4. Safety reports and safety audit reports (Rule 10) 5. Updating of reports under Rule 10 (Rule 11) 6. Requirement for further information to be sent to the authority (Rule 12) 7. Preparation of on-site emergency plan (Rule 13) 8. Preparation of off-site emergency plan (Rule 14) 9. Submission of data to authority for offsite plan (Rule 15) 10. Disclosure of information (Rule 16) 11. Collection, development and dissemination of information (Rule 17) 12. Import of hazardous chemicals (Rule 18)
EPPR Rules, 1996	EPPR Rules primarily deal with the constitution and functioning aspects of central, state, district and local crisis groups, assistance given to district and local crisis group along with information provision to public.

UNIVERSITY OF TWENTE.

Regulatory Framework

- ✓ *Operating mostly in the reactive mode, the regulatory system is often unable to pre-empt future or upcoming problems and the issue of management of environmental risk appears to be one of them.*
- ✓ *As a result, yet, at present there is no linkage of risk prevention with land use planning, an aspect which assumes significance considering the effect of the Bhopal disaster.*

UNIVERSITY OF TWENTE.

Regulatory Framework

- ✓ *Absence of national regulations on occupational safety and health and medical emergency management.*
- ✓ *Harmonization of classification and definitions in existing regulations including petroleum and petroleum products.*
- ✓ *Absence of regulations on storage and transportation of cryogenics*
- ✓ *Lack of legislation on risk assessment requirements and classification, labelling and packaging for industrial chemicals*
- ✓ *Need to identify technical competent authorities and standardization of reporting mechanisms for the implementation of various chemical disaster-related activities.*
- ✓ *Non-availability of statutes for grant of compensation to chemical accident victims.*

UNIVERSITY OF TWENTE.

Institutional Framework

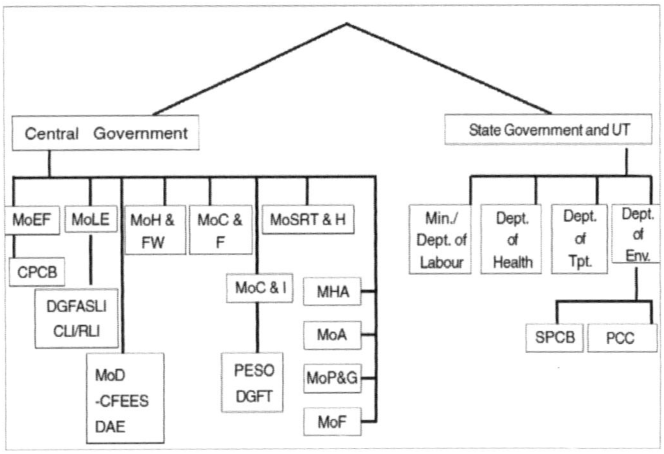

UNIVERSITY OF TWENTE.

Institutional Framework

- ✓ *Absence of an independent regulatory body (other than the competent authority) to assess the overall effectiveness of the system*
- ✓ *No structured and clear hierarchy to comprehend the institutional mechanism as well as the role of individual organizations*
- ✓ *There is often lack of both horizontal and vertical coordination between the agencies. In the vertical chain, there is more a tendency of central and state level agencies to get into a conflict.*
- ✓ *Lack of adequate institutional capacity at both the nodal (MoEF and CPCBs) and implementation level (SPCBs and others)*
- ✓ *At the horizontal level, lack of common database of information on the regulated community often result in duplicate efforts which result in industries to fill in multiple forms to meet the permitting requirements.*

ITC UNIVERSITY OF TWENTE.

Procedural Framework

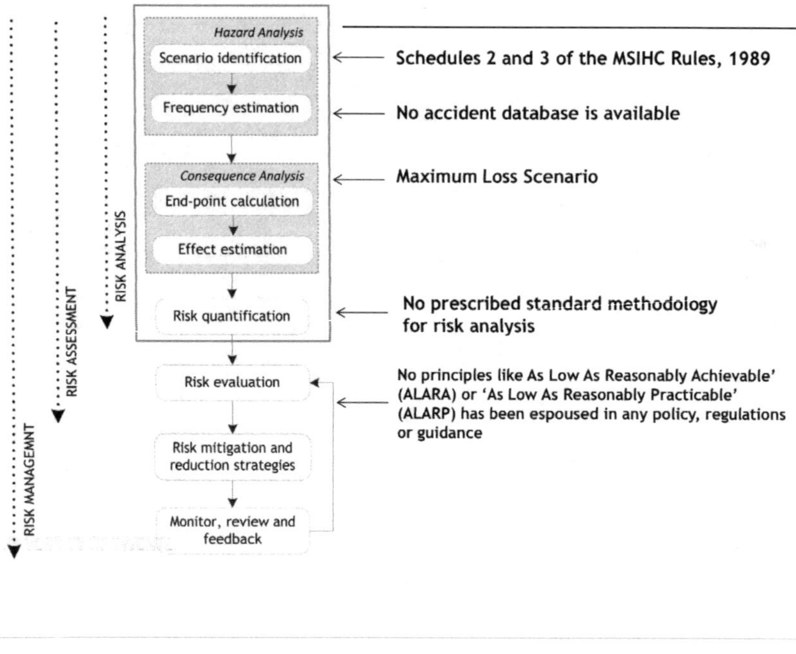

ITC

Procedural Framework

- *Lack of harmonized methodology for risk assessment*
- *Non-use of standard terminology*
- *No clear guideline for scenario selection (i.e. worse or most credible etc.) and criteria or limit to identify those scenarios*
- *Absence of yearly statistics of major industrial accidents for frequency estimation*
- *Lack of standardization to use models for QRA (e.g. weather condition, threshold limit)*
- *No standard damage criteria (i.e. fixed endpoints or Probit endpoints)*
- *No acceptable criteria (i.e. individual risk, societal risk, risk matrix) for risk evaluation*
- *Lack of cumulative risk assessment practice for industrial clusters*

UNIVERSITY OF TWENTE.

Thank you!

UNIVERSITY OF TWENTE.

Process Safety Management [PSM] Metrics

SHAKEEL KADRI shakeel@aiche.org

DIRECTOR, GLOBAL PROCESS SAFETY & RISK MANAGEMENT
AIR PRODUCTS AND CHEMICALS, INC
AND
CCPS EXECUTIVE DIRECTOR DESIGNATE

26 DECEMBER 2014
GLOBAL SUMMIT ON PROCESS SAFETY , MUMBAI, INDIA

Presentation Outline

- Process Safety Management [PSM] Metrics
- PSM Metrics Implementation at Air Products
- Leading Indicators – CCPS Project
- API/AFPM – Industry Analysis “Deep Dive”

Process Safety Management Metrics

- ✓ Leading Metrics – a forward looking set of metrics which indicate the performance of the key work processes, operating discipline, or layers of protection that prevent incidents.
- o Lagging Metrics – a retrospective set of metrics that are based on incidents that meet the threshold of severity that should be reported as part of the industry-wide process safety metric.
- ✓ Challenges to Safety Systems, Near Miss and other internal Lagging Metrics – Although these events lagging indicator events, they are generally considered to be a good indicator of conditions which could ultimately lead to a more severe incident.

API-754 PSM Metrics

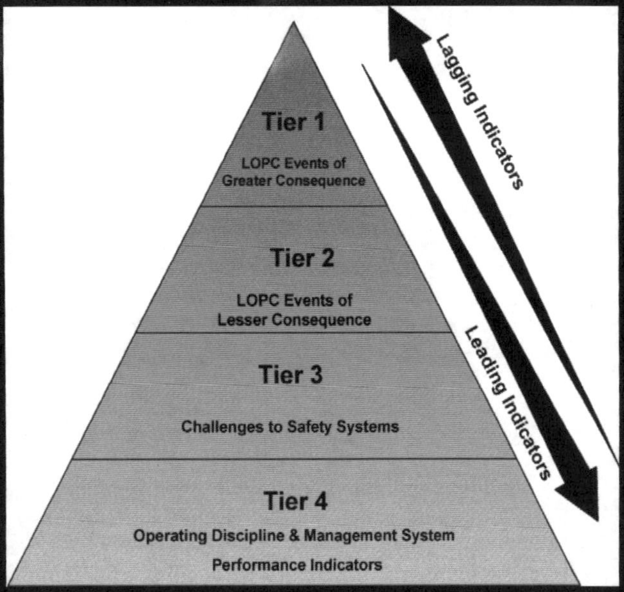

Air Products Related slides have been deleted from this presentation

Process Safety Leading Indicator Metrics

A CCPS Survey Project

Leading Indicator Metrics Project Team
Team Chair: Kenan Stevick

Author(s)

Sanjay	Choudhary	Tata Chemicals Ltd.	sgchoudhary@tatachemicals.com
Elroy	Christie	Honeywell	Elroy.Christie@Honeywell.com
Jeffrey	Fox	Dow Corning Corp.	jeff.fox@dowcorning.com
Eric	Freiburger	Praxair, Inc.	Eric_Freiburger@Praxair.com
R. Wayne	Garland	Eastman Chemical Co.	wgarland@eastman.com
Andrew	Goddard	Arkema	andrew.goddard@arkema.com
Shakeel H.	Kadri	Air Products & Chemicals, Inc.	KADRISH@airproducts.com
Darrell	Loewe	Formosa Plastics Corp., USA	DarrellL@ftpc.fpcusa.com
Steve	Marwitz	Formosa Plastics Corp., USA	SMarwitz@ftpc.fpcusa.com
Louisa A.	Nara	CCPS	louna@aiche.org
Tim	Overton	TOPS Consulting LLC	topsconsulting@hotmail.com
Jeff	Philiph	Monsanto	jeffrey.e.philiph@monsanto.com
Philip	Rasch	BASF	Philip.rasch@basf.com
Daniel E.	Sliva	CCPS	sliva@capital.net
Kenan	Stevick	Dow Chemical	KPStevick@dow.com

Presentation Outline

- Purpose / Intent
- Result Summary
- Data
- Leading Metrics Visibility
- Additional Guidance needed
- Barriers to Implementation
- Metrics Changed, New Metrics considered
- Next Steps

9TH GLOBAL CONGRESS ON PROCESS SAFETY

Purpose / Intent

- Purpose:
  - Provide an update on the use, direction, and effectiveness of Process Safety leading indicators in the chemical and petroleum industries.
- Strategic Alignment / Intent:
  - Leading Indicators are intended to help drive performance improvement and lead to reduction in the number and severity of process safety incidents. They do this by identifying improvement opportunities in the management systems for Layers of Preventative defense against Tier 1 and Tier 2 events.

Result Summary

- High percentage of companies recognize the value
- Industry is still 'experimenting and discovering' as to which indicators make the most sense
- Three different focus areas or approaches were identified as most effective in improving performance. These are:
  1. PSM Compliance
  2. Learning Experiences and Management of Deviations
  3. Management Engagement

Result Summary

1. PSM Compliance - Follow-up on actions across the spectrum of Process Safety Management Systems:
   - Audit Corrective Actions
   - PHA Actions
   - Completion of Safety Critical Equipment Inspections or Calibrations
   - MOC Actions
   - Unplanned Event Corrective or Preventive Actions

Result Summary

2. Learning Experiences and Management of Deviations
   - Process Safety Near Miss Reporting including fires
   - Challenges to Safety Systems in general and specifically calling out: Safety Instrumented Systems and Relief Device Activations

Result Summary [3]

3. Management Engagement
   - Picking the most pertinent measures to your operation and getting them in front of leadership;
   - Including Leading Indicators in agendas of various operational reviews and ensuring action.

Data Summary

- 43 companies responded to the survey; 41 Companies [95%] responded that they use leading indicators.
- The use of leading indicators varies by company, from a low of 3 leading indicators to as many as 28 leading indicators for a company.
- Each of the 25 leading indicators was used by one or more of the responding 41 companies.
- 20 companies [45% of responded to the survey] are using 12 or more leading indicators
  - 5 Tier 3 leading Indicators
  - 7 Tier 4 leading indicators

of Companies using a Specific Indicator

All 25 leading indicators were used by one or more of the responding 43 companies.

12 or more leading indicators were used by 20 or more of the 43 companies, (45%)

The red box on the chart highlights the 12 leading indicators used by the 20 or more companies

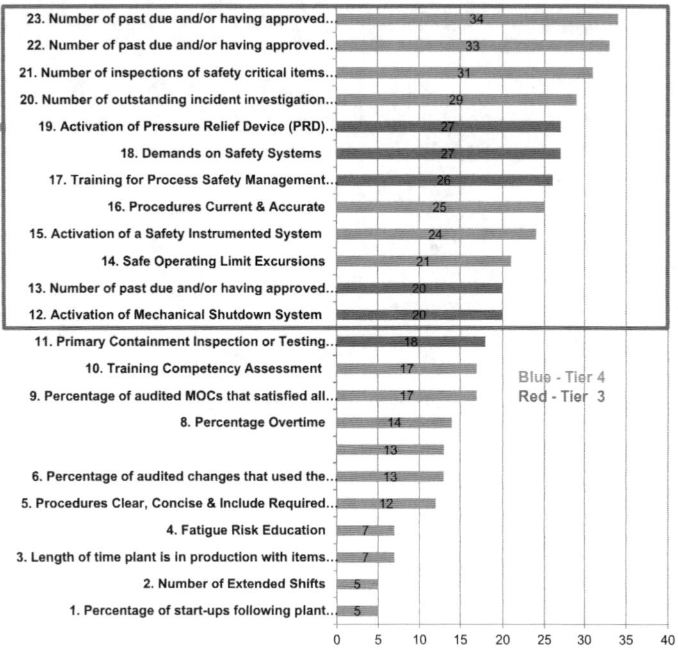

20 Companies or more Using these 12 Leading Indicators

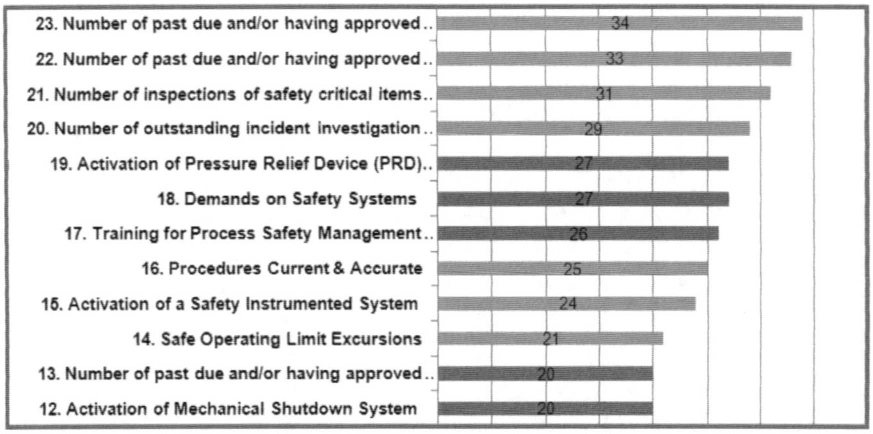

Frequently used Leading Indicators

Tier 3

1. Activation of Pressure Relief Device (PRD) Not Counted as a Process Safety Incident (PSI) or Loss of Primary Containment (LOPC)
2. Demands on Safety Systems
3. Activation of a Safety Instrumented System
4. Safe Operating Limit Excursions

Tier 4

1. Number of past due and/or having approved extension of audit action items
2. Number of past due and/or having approved extension of PHA action items
3. Number of inspections of safety critical items of plant and equipment due during the measurement period and completed on time
4. Number of outstanding incident investigation action items closed
5. Training for Process Safety Management (PSM) Critical Positions
6. Procedures Current & Accurate
7. Number of past due and/or having approved extension of regulatory issue

Leading Metrics Visibility

Most companies indicated:

- Roll their measures up into meaningful scorecards:
  - prompting management action,
  - review with senior leadership, (in some cases all the way up to board members)
- Publish data on internal review reports, websites and in newsletters for communication and action across the company

More Guidance Needed

- ✓ 15 companies felt that the existing guidance was sufficient.
- ✓ 8 companies indicated they had been using these metrics and were comfortable in their understanding.
- ✓ 3 companies indicated that they were not yet using these metrics.
- ✓ 7 companies felt additional guidance would be useful to ensure that "challenges to safety systems" are used consistently within and between companies.

Barriers to Implementation [1]

- Senior Management Commitment/support is essential for the implementation and sustainability of a successful metrics program.
- There can be differences in understanding metrics definitions across the company, e.g., different geographies, acquisitions
- Resources are needed in order to report metrics in a timely manner.
  - Maintaining trained resources who understood the definitions and how to extract the data from the computer tracking system presented a challenge due to transfers, turnovers and retirements.

Barriers to Implementation [2]

- Data Collection Systems
  - Data collection systems often do not readily produce the information companies want to track
  - Significant time and money are needed to develop information relating to leading indicators
  - Computerized / automated data collection systems are helpful in reporting in a timely manner
  - Communication to the workforce about the criteria for inclusion can take a great deal of time and effort
  - Employees need to hear that reporting is important, data consistency will not be perfect from the start but will improve over time

Metrics Change, Discarded and New Metrics being Considered

- 12 companies said they changed or discarded metrics
  - No longer represent areas where improvement needed
  - Emphasis changed to better align with underlying issue
  - Improved performance moved some to background
- 29 companies stated they were considering new metrics. Examples:
  - Alarm management
  - Senior leadership time in the field
  - Fatigue risk management
  - Process Safety Culture
  - Monitor procedures and Incident investigation quality

Next Steps

The following activities are scheduled to continue the focus on the development of leading indicators in the chemical industry:

- Publication of the White Paper in CCPS website for comments after it gets presented in the 9th GCPS
- Communication to CCPS Member companies to adopt leading indicators and inform CCPS of the list of indicators adopted along with the definitions and examples.
- The members of this Project Team to share their own company experiences in use of leading indicators.
- The review to be included in the 2014 Global Congress on Process Safety.
- Biannual Survey from the CCPS members

9TH GLOBAL CONGRESS ON PROCESS SAFETY

API related slides have been deleted from this presentation

shakeel@aiche.org

QUESTIONS??

UNEP

Process Safety Management within the Post-2015 Sustainable Development Agenda

Dr. Isabelle Louis
Regional Deputy Director for Asia and the Pacific
United Nations Environment Programme (UNEP)

Overview

- United Nations Environment Programme
- Environmental Trends, Risks and Opportunities
- Trends in Chemical Production and Use
- Sustainable Development beyond 2015
- Industrial Risks and Sustainable Development
- UNEP's Response- Safer Production
- Mainstreaming Safer Production- Importance of Partnerships

UNEP

UNEP Programme of Work

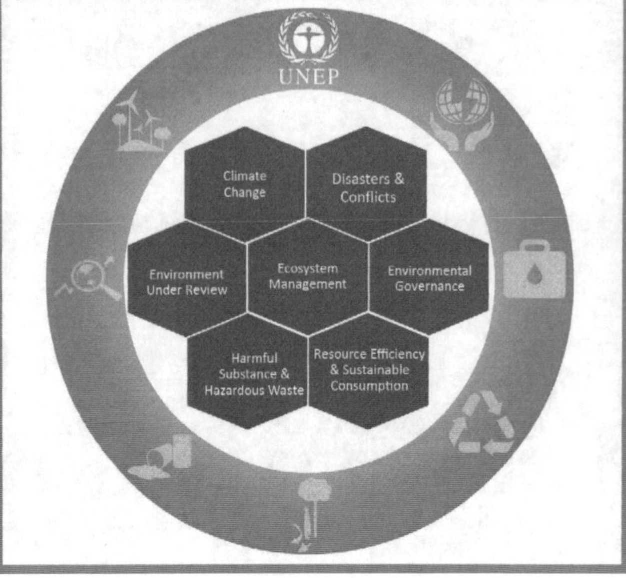

UNEP

UNEP and Industry

- Catalyses a transition towards **low carbon, resource efficient** and equitable development based on the **protection and sustainable use of ecosystem services, coherent environmental governance** and the **reduction of environmental risks** for the well-being of current and future generations in order to contribute to sustainable development.

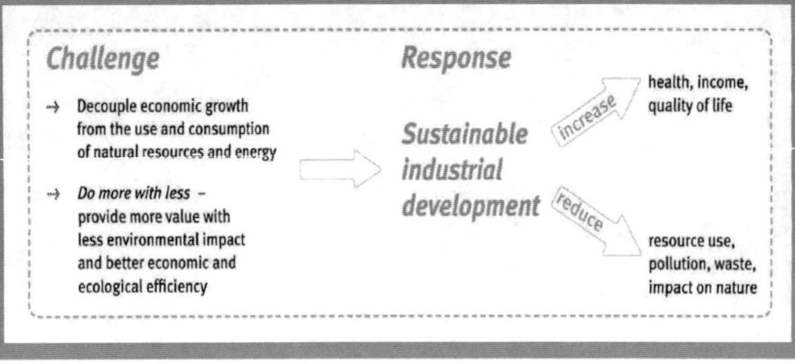

UNEP

UNEP and Chemicals Management

- **Strategic Approach to International Chemicals Management:**
  - A policy framework promoting chemical safety globally
- Objective: **To achieve sound management of chemicals throughout their lifecycle so that, by <u>2020</u>, chemicals are produced and used in ways that minimize significant adverse impacts on human health and the environment.**
- Framework emphasizes chemical safety as a sustainability issue
  - Covers risk reduction, knowledge and information, governance, capacity and illegal international traffic
- Global Plan of Action item on emergency prevention and response measures

UNEP

Environmental Trends

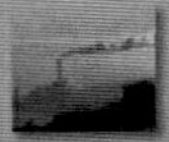
Atmosphere
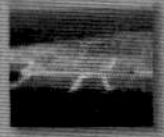
Land

Water

Biodiversity

Chem / Waste

- GHG emissions to double in next 50 years
- Urban uses require 100-200 million hectares of land over next 40 years
- Alarming water scarcity affecting people in some regions
- 230% rise in flood disasters (1980-2000s)
- Critical habitats continue to be endangered
- Lack of data to document risk to human health and environment from chemical products

UNEP

Environmental trends pose risks and opportunities

230% rise in flood disasters → Operational/supply chain disruptions

38% rise in drought disasters 1980-2009 → Increased demand for resilient products

Global water withdrawals tripled in last 50 years → Constraints on growth; New markets for water treatment, efficiency

20-50 million tons of e-waste generated annually → Customer/regulatory pressure to reduce waste; Growing market for waste recovery

UNEP

Trends in Chemical Sector

UNEP Global Chemicals Outlook:

- Virtually all manufacturing processes rely on chemicals
- Since 1970s global chemical production has been shifting from OECD countries.
- Ca. 80% of new chemical production capacity in emerging economies
- Significant chemical production growth rates are expected until 2020:
  - 46% growth rate in the Asia Pacific Region
  - 40% in Africa/Middle East
  - 33% in Latin America

UNEP

Trends in Chemical Sector
GEO-5 for Business (2013):
RISKS
OPPORTUNITIES
Operational
• Rising energy prices
• Water scarcity
• Business continuity
Diversification of feedstock away from fossil fuels
Market
• Consumer and community concerns about environmental and health impacts reduce market demand for chemicals
Larger markets for specialized chemical sector products, water treatment technologies, agricultural inputs
Policy
• Product use restrictions and phase outs
Green chemistry, chemical alternatives and substitutes and chemical leasing
Reputation
• Increased community sensitivity to the potential impacts of chemical plants
Competitive advantage from green chemistry

Trends in Extractives Sector
GEO-5 for Business (2013)
RISKS
OPPORTUNITIES
Operational
•Extreme weather- damage to infrastructure
Increased access to resources
Market
•Reduced demand for virgin minerals in favour of recycled materials
Increased markets for cleaner fuels, and for recycled and more sustainable minerals
Policy
•Regulatory limits on mercury, air emissions and water discharges, and conservation of biodiversity
Climate policies driving a change towards natural gas
Reputation
•Reputational damage and potential loss of social license to operate
Partner in solutions for climate change

Vision Post-2015

- Year 2015 is pivotal for climate change, disasters and sustainable development.
- New international frameworks:
  - Framework on Disaster Risk Reduction, March 2015
  - Sustainable Development Goals, September 2015
  - Climate Agreement, December 2015
- Unique opportunity for action by governments, private sector, civil societies and international organisations
- Encourages new partnerships for implementation

UNEP

Sustainable Development Goals

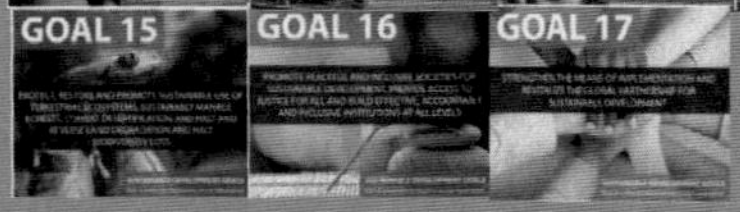

- Goal 9: Build Resilient Infrastructure, Promote Inclusive and Sustainable Industrialization and Foster Innovation
- Goal 11: Make Cities and Human Settlements Inclusive, Safe, Resilient and Sustainable
- Goal 12: Ensure Sustainable Consumption and Production Patterns

UNEP

Industrial Safety

- **Embedding safety into the operations of industry is vital for Sustainable Industrial Development. Risks however include:**
  - Social: Immediate injury, short and long term health problems, death, infrastructure loss, medical costs
  - Environmental: Loss of Eco-system services and natural resources
  - Economic: Loss of workplaces, income and livelihoods, costs of rebuilding
- **Developing countries are often at a greater risk due to lack of regulation, enforcement and limited resources**

UNEP

Industrial Safety

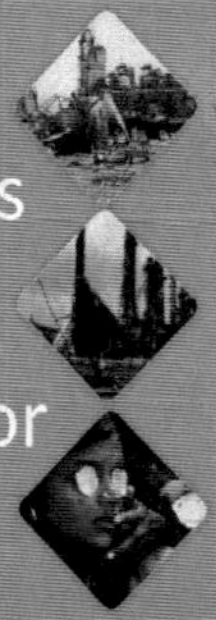

Emerging considerations:

- Intensification of extreme weather events pose new risk for industrial safety
- Scale of urbanization poses a challenge for industrial/residential land-use planning
- Significant number of SMEs involved in industrial operations across supply chains

UNEP

Zero Accidents

"If you think safety is expensive, try an accident..."

"Good Safety is good business"

- SMEs form the backbone of economies, but often lack an understanding of risk, best practices and business case of risk management
- An accident can ruin an SME and can have widespread consequences to other companies

- Professional and technical organisations, academia, and large companies can play a significant role in raising awareness.

UNEP

UNEP's Response- Safer Production

- **UNEP Governing Council requested UNEP in 1986 to work in partnership with governments, industry and other UN bodies to establish a programme that enables multi-stakeholder cooperation to minimise risks**
  - **Led to the establishment of the APELL programme**

***Ultimate goal**: Reduce the consequences (on health and environment) and the likelihood of chemical accidents*

- **UNEP supports and collaborates with governments, private sector and civil society on awareness, prevention, preparedness and management of industrial risks/chemical accidents**

UNEP

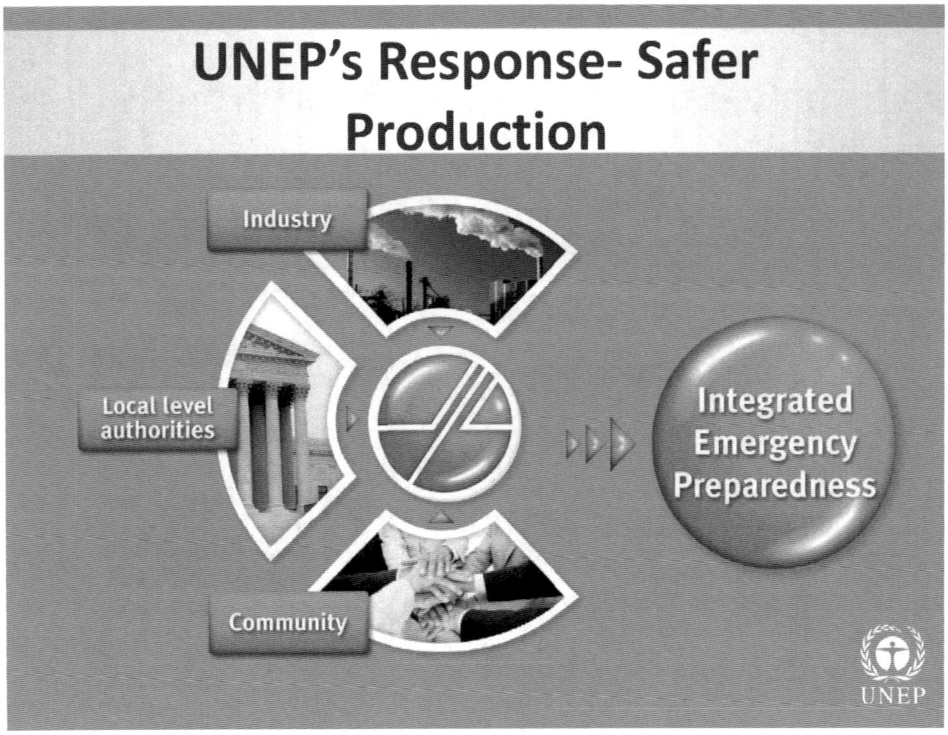
UNEP's Response- Safer Production
Industry
Local level authorities
Community
Integrated Emergency Preparedness
UNEP

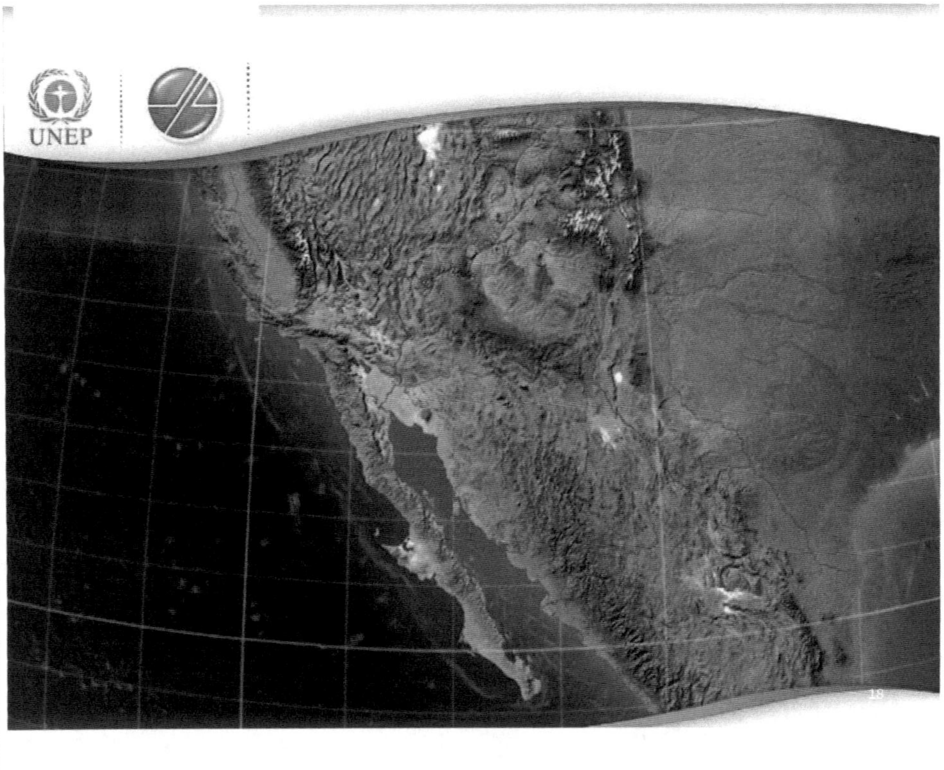
UNEP

UNEP's Response- Safer Production

Today UNEP's integrated approach to industrial risk reduction has three pillars:

I. Strategic guidance and support to national governments in establishing legal systems, policies and programmes on chemical accident prevention & preparedness
II. Facilitation of APELL processes worldwide
III. Reaching out to SMEs through national technical service providers

Mainstreaming Safer Production- Partnerships matter

Ensure Sustainable Consumption and Production

Includes *inter alia*:

- Sound management of chemicals
- Mainstreaming **sustainable consumption and production practices by companies**

To achieve this, industry is challenged to prioritise:

- Resource efficiency
- Cleaner and **Safer production**
- Corporate Social Responsibility

Mainstreaming Safer Production – Partnerships matter

- **Policy Level**
  - Mainstreaming process safety at the policy level as part of disaster risk reduction
  - Enacting programmes and policies ensuring effective risk management and reducing creation of new risk
- **Company Level**
  - Mainstreaming process safety in business operations
  - Prevention of new risk- risk sensitive investment
  - Enhancing technical capacities to operationalize global principles in operations- including SMEs

UNEP

Mainstreaming Safer Production- Partnerships matter

- **Industrial Safety is Everyone's Business: Partnerships are key!**
- Significant opportunities for organisations to work together and to establish new kinds of partnerships:
  - Private public
  - Across sectors and countries (north-south, south-south)
  - In supply chains
- **UNEP sharing its experience on 'Responsible Collaboration' :**
  - **Establishing multi-stakeholder processes for enhanced cooperation**
  - **Sharing lessons from accidents**
  - **Working with SMEs in the supply chain**

UNEP

Mainstreaming Safer Production-Partnerships matter

- CCPS Vision 2020: Recognises societal role in striving for process safety excellence

- Opportunity for a private sector leadership role in establishing partnership:
  - Multi-stakeholder collaboration through APELL
  - New partnerships for collaboration established
  - Networks for sharing best practice

There is a great opportunity accelerate the ambitions set by vision 2020 and to make progress in Sustainable Development

UNEP

UNEP

THANK YOU!

Mechanism and Platforms to ensure Meticulous Verification of PSM Implementation at RIL

VISHAL PATEL

RELIANCE INDUSTRIES LIMITED,
JAMNAGAR MANUFACUTING DIVISION
JAMNAGAR
Vishal.Patel@ril.com

Prepared for Presentation at CCPS Global Summit on Process Safety
15th & 16th December, 2014

Introduction

Organizations involved in managing the process risks are having opportunities where in they can review/assess their performances and take appropriate steps to improve it. Various verification methodologies are already being adopted by the Organizations and are being practiced to ensure that Process Safety Management (PSM) systems are working as intended. However, it is important that these verification methodologies point out the deficiencies in the Management System before they can lead to maximum credible loss scenarios.

In Vision 20/20, released by CCPS, such Performance Assessment by Industries have been defined as one of the four societal themes which is named as 'Meticulous Verification'. The concept of this verification includes participation by Third Parties, Public and Non-Governmental Organizations (NGOs) along with Industry's own verification platforms.

Challenges at RIL

At Reliance Industries Limited (RIL), the efforts of structured PSM Implementation started way back in 2007 under 'Radical Complete Transformation' project. Across the Organization multitude of efforts were initiated for formation of Standards, Procedures and conducting training at all levels under guidance of M/s DuPont Safety Resources.

On achieving certain level of maturity, the challenge then was, 'How to Verify' that the performance achieved is in line with the, Organization's Objectives, and meets the expectations & targets as specified in Corporate standards. Once the verification mechanism was framed, the next level of challenge was to effectively use the data reported by various mechanisms.

This paper briefs about the platforms and mechanisms established for meticulous verification of PSM Metrics & it also describes the efforts initiated by RIL on each of these fronts.

1 PSM Metrics: Measurement, Monitoring and Its Usage

1.1 Defining the proper set of metrics

Defining the proper set of metrics is the first step towards effective PSM implementation. If the performance is not 'measured', then it can't be 'improved'. However, these metrics shall be 'Dynamic' instead of 'Standstill' so that they match with the maturity of PSM Implementation within organization. Let's take the following example for the same.

One of the requirements under PSM – Process Technology Element is availability and accessibility of Hazard related information for all chemicals handled at facility. At the initial level of PSM implementation, Organization may focus on having Material Safety Data Sheets (MSDS) available for all chemicals and its accessibility to all employees through intranet or hard copies. The same can be defined as Metrics e.g. Number or No. of Chemicals Handled Vs. No of Chemicals' MSDS uploaded on intranet. On achieving maturity on this metrics, the focus can be shifted to the MSDS quality and the appropriateness of the data contained in it, e.g. 16 point format MSDS instead of 8 point format MSDS which defines the hazards related properties in the detailed manner. Such gradual improvement steps can help in stepping upwards on the ladder of PSM implementation.

At RIL, we have a predefined set of PSM Metrics and these were defined right from the beginning of PSM Implementation. Two rounds of reviews have been conducted so far, the Metrics where the maturity of implementation is achieved have been dropped and the next levels of metrics have been added based on PSM implementation Level.

1.2 Institutionalizing the mechanism for Metrics reporting and review

Once the set of metrics are defined by the Organization, structure and methodology are to be defined for its reporting. This structure shall include 'Who will compile the data', 'Who will report?', 'To whom the metrics are to be sent'?

The document defining the metrics shall also define the source of metrics, frequency of measurement and unit of measurement. Both the reporter and receiver of the metrics are to be trained on 'How one should compile the data?' and 'How they should interpret the data?'

As discussed in earlier section, metrics shall match the level of implementation and at the same time it shall have the multi-tier approach for its reporting. Such approach can be metrics reporting at Plant Level out of which certain critical metrics parameters can be included for Site Level monitoring and further to Group Level monitoring.

This will help in resolving the issues at various level of management and will ensure appropriate focus also. Reporting all metrics at all above defined levels may lose its intended outcomes.

A simple figure describing the level of review for the PSM Metrics at Site Level, Group Level and Board Level is shown below for reference:

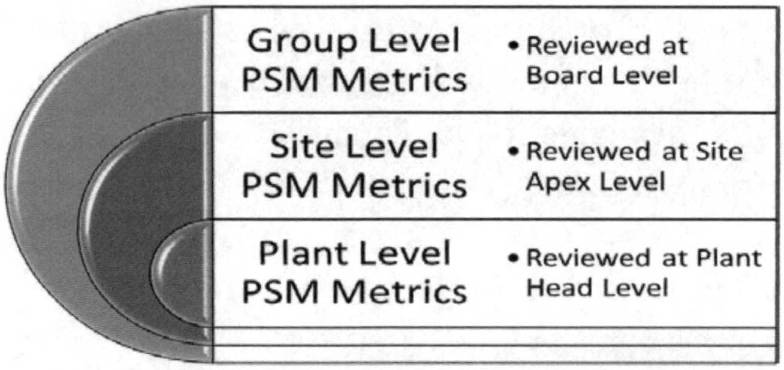

Figure – 1: Level of Review of PSM Metrics

At RIL, there are defined set of metrics at plant level which are reported by Plant PSM Coordinators in consultation with their Plant Heads. These metrics are being compiled at Site Level by Site PSM coordinator. Site in turn reports the critical metrics at Group Level to Group PSM Competency Lead. Critical PSM Metrics are reported by Group HSE Dept. to HSE Council which involves Board Level Members as its Chairman.

1.3 Analyzing the PSM metrics trends

Once the Metrics are set and the reporting structures and mechanisms are designed, it is very important to analyze and interpret the data reported. It is also important that such analysis is presented to the Top Management wherever, their intervention is deemed necessary.

At Reliance, we have started a practice of discussing the PSM Leading Indicator Metrics on monthly basis during PSM Performance Review Meetings where Site Apex Level members are also present. Appropriate decisions are being taken on proactive basis against negative trends as well as the efforts to enhance and sustain the achieved positive trends are discussed in these meetings.

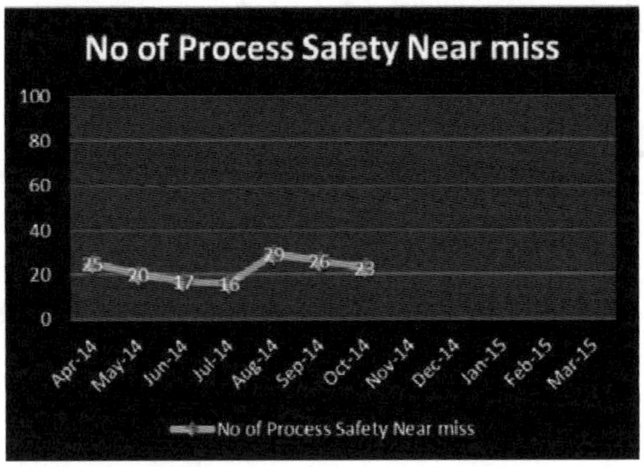

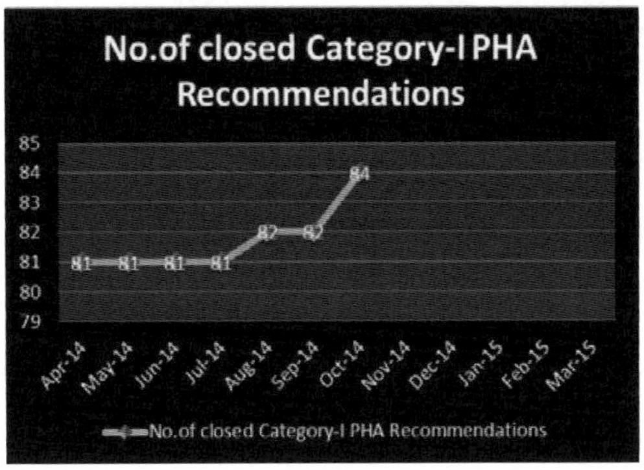

Figure 2 – Trends and Analysis of PSM Metrics

2 Audits:

2.1 Developing PSM Audit Program

Audit is very important tools for verification of PSM Performance. Especially PSM Audits can play vital role in improving the performance of an Organization. It is very essential that enough emphasis is given in deciding the Audit Program at Organization.

At, RIL we have established a 'Three Tier' Audit system as described below

1st Party Audits:	Conducted at Inter Plant Level by Site's own Internal Auditors
2nd Party Audits:	Conducted at Intra Site Level by other Site's Auditors
3rd Party Audits:	Conducted by Third Parties

Once the audit mechanism is decided, then the frequency of audits and audit team members composition are also to be defined in Group Standards and Procedures. Internal audits require development of pool of internal auditors, who can be certified through Internal Trainings and Competence Development Programs.

Once the frequency and Team Composition are decided, the next task is to define the Audit Protocols. Defining Audit Protocols helps in safeguarding uniformity across all audits being conducted by all auditors. At RIL, we already have set of predefined audit protocols for PSM audits, which define the RIL standard requirement in terms of questions in checklist form. These protocol questions are designed for each of the PSM elements. Like PSM Metrics, it is also important to keep these audit protocol questions up to date.

Once the audits are performed, it is very important to comply with the recommendations of the audit findings until corrective and preventive actions are implemented. Trends of audit findings as well as the repeat findings are the important indicators which shall be presented to Site Top Management as a part of review mechanism. This will help in eliminating the repeat findings across the Site as well as ensure Management Commitment and intervention in resolving site wide issues

At RIL, all the findings of PSM Audits are being entered against relevant action parties, in online 'Corrective Action Tracking Portal' with clear, specific and time bound action plan. The closure trends as well as long pending audit findings are reviewed in Monthly Site PSM Review Meetings.

2.2 Effectiveness of PSM Audits

Achieving effectiveness of the PSM Audits is a big challenge for an Organization whilst implementing the Audit Management Program.

The audits shall not be shallow and review the 'Superficial Compliance'; however, they shall be of Deep Diving type so as to ensure that the system related gaps are identified instead of pointing non compliances. Audits which are conducted with the system improvement focus can help in trimming the deficiencies at their root level and in turn prevent is recurrence.

Audits shall be 'Conformance' based rather than 'Compliance' based. The Conformance Based Audits verifies / identifies the pitfalls / gaps of the PSM system. Below mentioned table describes the difference between these two types of audits:

Table 1 – Difference b/w Compliance Based Audits & Conformance Based Audits

	Compliance Based Audits	**Conformance Based Audit**

Focus Area	Checks compliance status with respect to organizations' laid down standards	Identifies Weakness of the plant/site in maintaining the safeguards identified for process risk scenarios
Methodology	Checks documents, verifies records (hard copies , soft copies) for checking the compliance status	Verifies effectiveness of the safeguards , awareness on importance of maintaining safeguards at shop floor level

At RIL, as a part of continual improvement, we have initiated actions for changing the focus area of the audits from compliance based to conformance based. A number of efforts are initiated in developing the Risk Register at various levels of Organization so that identified safeguards and controls can be checked effectively.

3 Digital Technology Usage

Use of Technology benefits effective verification of Process Safety Parameters before they can contribute to 'Loss of Containment' of Materials. Following are some of the tools implemented at RIL using such Technological aids.

3.1 Integrity Operating Window (API 584)

This concept ensures that the process parameters are maintained well below the design limits and any excursion of the parameters are analyzed with its corrective actions. Normally the process parameters are maintained within operating envelopes which are much below the design envelopes. Integrity Operating Window (IOW) limits are defined in between these two envelopes through digitized software. During plant operations any excursion is being captured by this system, and based on level of excursion alert mails are being sent through system. During the daily meeting these excursions are being reviewed with its possible cause and corrective actions are identified to prevent such excursions. The monthly IOW trends are regularly being reviewed at RIL and if there are consistent excursions, plans are defined for its corrective actions.

3.2 Integrated Operation Management (IOM)

At RIL, following methodology is used in structured manner in our journey towards Operational Excellence:

- Effective Hand Over Take Over: At the end of any shift, outgoing Shift Superintendent logs the key list of jobs, activities and trips encountered out by him during his/her shift in Online portal. The incoming shift has to log in and accept the same after reviewing and plan accordingly his/her shift. This benefits in effective communication and understanding of process safety risks on a continuous basis.

- Shift Team Meetings: Across our Organization Shift Team Meetings are organized where in at the start of any shift, all the key shift personnel across the plant interact with each other through the Online Portal using Video Conferencing / Physical Meetings depending on feasibility so that, they can share their individual plans, conflicting job activities and share resources for safe completion of jobs where key process safety risks are to be dealt with.
- Operator Driven Reliability: Plant Operators carry out proactive monitoring of key PSM Critical equipment through structured operator's rounds, where in they are given "Hand Held Device" which are designed for recording data which will be later analyzed by them for enhancing safe operating conditions of various processes time to time.
- Empowering operators, leading to effective management of abnormal situations, "Stabilize, Slow down or Shut-down if necessary".

A lot of efforts have been put in by our Organization in training people for using the above tools, in cultivating the culture of commitments and accountability for taking process safety ownership.

4 Verification by External Agencies / Third Parties

Verification by External Agencies is very vital as they assess the Organization's system with an independent eye. Such independent reviews help organization to reveal some of the identified process safety risks.

At RIL, renowned External Agencies from around the World were invited to asses our PSM Safety. In the past, following external agencies carried out a number of reviews at our Organization:

- Risk Management Review
- Alarm Rationalization Reviews
- Ensure Safe Production Review
- Safety Management System Evaluation
- Safety Management System 'Five Star' Audit
- Review of Chlorine Safety Handling Systems
- Emergency Preparedness Review

The improvement opportunities identified in all these above mentioned reviews are being tracked till its appropriate closure. This helps the Organization in continually taking forward the system to the next level.

Other than reviews, Benchmarking is also one of the important tools in calibrating the performance of Organization with respect to 'Best Performing' Organizations at Global or National Level. Such Benchmarks will identify the new horizons for improving Organizations Performance.

At RIL, we have participated in following Benchmark studies

- Process Safety Benchmark
- Benchmark studies of Refineries

5 Verification by Government Agencies :

Verification by Government Agencies can play a pivotal role in ensuring the conformance to the safety requirements essential to achieve good PSM performance. Such verifications shall go 'Beyond Legal compliance' and if necessary existing legal framework is to be updated to facilitate more rigorous and effective verifications.

Government agencies can also play a role in providing the 'Best Practice' sharing platforms across the sector so that lessons learnt and good practices can be shared and implemented among all the Industries.

At RIL, we have undergone following verifications conducted by Oil Industry Safety Directorate (OISD) during the last couple of years. These assessments have brought out a number of actions which were really helpful in strengthening our process safety safeguards. To name a few:

- External Safety Audits (ESA)
- Surprise Safety Audits (SSA)

6 Future State Plans

In spite of all efforts mentioned above, we at Reliance have not felt it enough for an effective process safety performance. We are continuously striving towards bettering our performance. Below mentioned are examples of our future state plans for effective verification mechanism for our process safety performance.

6.1 Metrics on Dashboards

At our organization, currently PSM Metrics are being reported by plant and site level through manual Excel worksheets. We have plans for automating the reporting, trending and analysis part. Not only automating, we are also planning to develop position wise online dashboards in which critical PSM Metrics parameters (based on relevance) will be displayed online, such dashboards will display real time data also wherever they can be linked with the real time data loggers.

6.2 Smart Procedures

As a part of 'Project Health Check' exercise, we realized that our Operating Procedures (both Standard Operating and Standard Maintenance) require enhancement. Post health check group wide drive was taken for strengthening these procedures. Cautions, Alerts, Hazard Related Information of the activity being carried out were added in all these procedures. Diagrams and linkage with the other procedures was also ensured during these enhancement exercises.
However, still we want to covert our procedures to smart procedures with an automation where any updates can be done online and sent across up to shop floor. Any improvement suggestions can be logged online and tracked till its resolution.

7 Conclusion

Building a sustainable Process Safety Management System shall be a moral responsibility for any Organization.

A number of critical PSM Elements were addressed in this paper. If systematic efforts are devoted for measuring and verification of identified PSM Metrics, then good process safety performance can be achieved.

Process Safety Management System acts as a bridge in the journey of an Organization towards Operational Excellence and at the same time it ensures an injury free, safe and environment friendly operations which benefits the organization society and community at large.

CCPS GLOBAL SUMMIT ON PROCESS SAFETY

An AIChE Technology Alliance
CCPS
Center for Chemical Process Safety

Enhanced Safety from Advanced Simplicity

CCPS – AIChE Global Summit on Process Safety

December 2014, Mumbai

Stanley Joseph

Contents

Enhanced Safety from Advanced Simplicity

A challenge safety system designers recurrently confront is to come to a decision on the optimal architecture for their "safety loops". The good news is that there are well recognized and proven schemes that can help "compute" relevant metrics to enable choosing the best structure based on predetermined Safety and Reliability Targets.

When it comes to selecting the optimal "safety system" though, the decision-makers and Plant Managers are not always equipped with pertinent information for differential analysis and often have to rely on marketing information promulgated by safety system manufacturers.

A fundamental topic with regards to Safety System that is shrouded in some vagueness and obscurity is that of Safety System Internal Architecture. It must be emphasized that this topic should not in itself be something that Safety System Users really need to look into or be concerned about as long as they view Safety Systems from the correct safety performance perspective or do not specify any particular "internal" system architectures.

The reality, however, is that many organizations still stipulate specific internal architecture (CPU) in their Safety Instrumented System Requirements Specification.

Specification – a science or an art or both.

In order for any Requirements Specification to be complete and explicit to the accurate needs, it should appropriately specify the behavior or characteristics that are relevant to the higher level objective. It is also equally important to avoid specifying any internal structure, design or realization method unless it is recognized as a constraint.

Now, let's visualize a situation that for a particular level measurement within a high pressure, corrosive and low density application / environment, a certain type of rugged, multi-switch point Magnetic Level Switch was considered optimal, 15 years ago. Locking-in that specific technology as part of future instrumentation specification for the same application would result in not benefitting from the immense progress (or even lessons learnt from technological pitfalls) that has taken place in the field of instrumentation, communication, calibration and smart devices since then. Worse still this can increase dependency on obsolete technology in related fields such as communication and interface management. It is evident that to specify the pertinent objective, process specifics, metrics and the environment would be much better than to specify the instrument or the technology itself.

Enhanced Safety from Advanced Simplicity

With the first generation of Safety Systems, Hardware majority voting was the only way to out-vote an error in one component. The practice of specifying internal safety system architecture possibly would have been appropriate in an era when Safety Systems were not the norm and even application-centric computerized Control Systems were yet to gain full acceptance in industry.

It is not seldom that one comes across User's Safety Requirements Specifications that specify system architectures from earlier generations of Safety Systems (TMR, DMR, and Quad), which employed and relied heavily on "hardware" redundancy and "hardware" voting techniques to enhance safety and availability).

This Safety Instrumented System qualification based on hardware architecture, voting structure etc can be not only distractive but is also disingenuous, considering that developments in technology currently afford the capability to run independent functions using the same hardware and software to SIL3 Integrity. That is to say, SIL3 Safety Functionality can be realized using even a common controller for Safety and Control application. The relevant safety metrics are Achievable Safety Integrity, Safety Availability and in particular, in my personal opinion **the ability to support preservation and continuance of its Safety Integrity through its life-cycle**.

Requirements should be oriented around the needs and should be an accurate elaboration of a higher-level objective - not oriented around how we achieve the desired goal, target or result**,** which in the case of a safety Instrumented system should be the safety performance, and may include aspects such as maintainability etc. In short, specify the needs not the means.

Can redundancy be a good thing?

Redundancy can be perceived simply as the addition of resources beyond (the bare minimum requirement) that is needed for the normal system operation. Redundancy was the best engineering solution that was developed in the earlier generation system for safety and availability. It is clear that Redundancy can help ease Dependency.

Hardware redundancy is the addition of extra hardware, usually for the purpose either detecting or tolerating faults. Hardware redundancy also allows for hardware voting. Hardware voting typically amounts to static techniques that rely strictly on fault masking and therefore results in degraded mode of operation when a fault occurs. These were, however, excellent techniques that lend itself to critical safety and reliability applications.

We can also have too much of redundancy and like all good things, too much of it can counter safety, especially if redundancy is applied without diversity. More hardware could effectively translate into more random failures (attributable to redundant hardware), leading to increased number of correctable errors and associated maintenance requirements - redundant components are only as dependable as their upkeep.

Setbacks can include increased weight, increased power consumption and increased cost. More than a single "dangerous undetected" failure of voted hardware components (for example in 2oo3 configuration, HFT=1) can put the safety system into an unsafe condition – unable to respond to a real process trip demand.

Modern safety system are lean on hardware but employs the advances made in microprocessor technologies that have made it possible to analyze relatively massive amounts of data, in real time, thus allowing for self-diagnosis, fault-detection and even correction resulting in improved safety systems. These remedial or counteractive measures were not practicable during the first generations of Safety systems.

Increased Diagnostics from Software and Time Redundancy

The prime requirement of higher safety integrity combined with the need for high availability of plants have led to considerable research and development efforts in the field of overall safety and control engineering, in the past decades. The areas of special interest have been the development of fast acting emergency shutdown system, on-line fault detection facility for the protection circuits, enhanced instrumentation capability for measurement of critical plant parameters and computerized systems for plant control, and protection.

The safety strategy adopted and the consequent development of special safety systems combined with extensive use of microprocessors in safety related instrumentation has allowed the distribution of processing capability even to the peripheral

3rd Generation Safety Systems such as SIMATIC S7-400FH, configurable to be fully redundant is based on a simple and very high Safety-Availability architecture 2(1oo1D) - meets SIL3 requirement with a single controller. This is an example of the leanest architecture with the maximum availability.

The 2(1oo1+D) architecture employs the concept of:

- Software Redundancy and
- Time Redundancy

Software redundancy is the addition of extra software, beyond what is needed to perform the given function of the safety trip. This allows for extra information (diagnostics and system status) beyond the logic itself; for example, error detection codes to detect and possibly tolerate faults.

Enhanced Safety from Advanced Simplicity

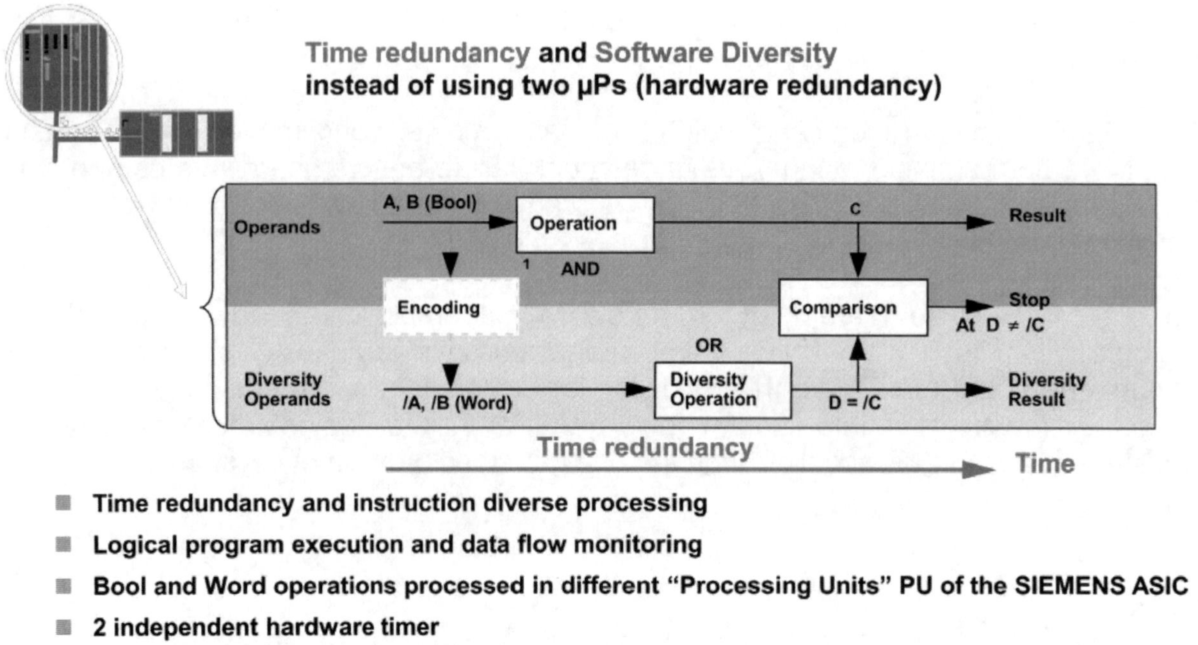

Time redundancy uses additional computations repeated at different points in time and then compared to perform the functions of a system such that fault detection and often fault tolerance can be achieved. Transient faults are also tolerated by this.

- During the first computation, the operands are used as presented.
- During the second computation, the operands are recomputed with swapped operands. This provides software Diversity.

The use of encoding function is made so as to allow faults in the hardware to be detected. No extra hardware is required but very high level of Diagnostics can be achieved.

Safety Life-cycle Management for prevention, protection and mitigation intent

Studies have shown that close to three-fifth of control system incidents / accidents were caused by oversight or inaccuracies made in the specification, design and implementation of the project. Furthermore, one-in-five of control system accidents were attributed to mistakes made in the course of "**changes after commissioning**".

The above figures suggest that in almost **eighty percent** of these accidents, even if the control or safety system responded as designed and implemented, it **would not have**

averted the **mishap**. Where then is the safety gap? Was safety lost in translation from requirements to implementation ?

Perhaps, the process analysis did not capture the scenario(s) that caused the incident. Worse still, conceivably the design did not successfully incorporate the safety functions identified in the Analysis Phase and the resulting Safety Requirement Specification (SRS).

One critical area that is easy to fail but often seems to be not receiving as much attention as necessary is the Safety coding / programming component.

The vulnerability that the program may not indeed be entirely and accurately translated from the SRS is compounded by the fact that application programs are **seldom validated by specialist third party**. Some aspects of Safety app-software might not be covered comprehensively in the SRS. Ensuring that the software is fail-safe then becomes the responsibility of the programmer. Acceptance Tests cannot always uncover consequential "Systematic" flaws.

Standards for process safety have recognized this problem and require that the implementer of all phases of the Safety Life Cycle are adequately trained and qualified to undertake duty for part of Safety Life Cycle assigned to them. As an Independent Safety Auditor (ISA) Consultant, I have come across non-fail-safe programming in segments of Safety Systems, every so often due to oversight – not as a result of lack of training and familiarity.

Unfortunately, these defects may not always be detected or corrected during Acceptance Tests. Even experienced Safety System programmers approach certain crucial safety-related functions such as over-rides and bypasses differently. That is to be expected seeing that there is more than one way of reaching the same outcome. The problem is that when a program is modified by someone other than the original author (which is almost always the case, for "**changes after commissioning**") there is potential for covert errors beings introduced, reducing the SIS Integrity. Translation errors infiltrate across multiple phases in the Safety Life Cycle. Qualification or experience of the responsible person may not be sufficient to prevent these costly slip-ups.

There is not only the need for **applying good Management of Change** (MOC) practices but also the need for **safety systems that facilitates MOC**. It is unrealistic to expect that a program that was once tested at FAT / SAT or commissioning not needing to change, in due course OR to expect the person making changes later to fully understand and incorporate all elements as done by the original author, at the time of design.

Additionally over a period of time, after the application software has been worked-on by multiple persons, it turns into a collection of signature-schemes from the personalities who contributed to the safety program in the course of its Life-cycle. As an author and auditor of safety system software, I can testify that it can be a recipe for harsh adversity.

"Simplicity is the ultimate sophistication"

A well-developed Safety Requirements Specification with detailed cause and effects chart is clearly a good way to organize and document the plant's safety strategy. C&E charts can be successfully even for sequence control. It is possible to simplify and represent complex sequences into Cause and Effect Matrix.

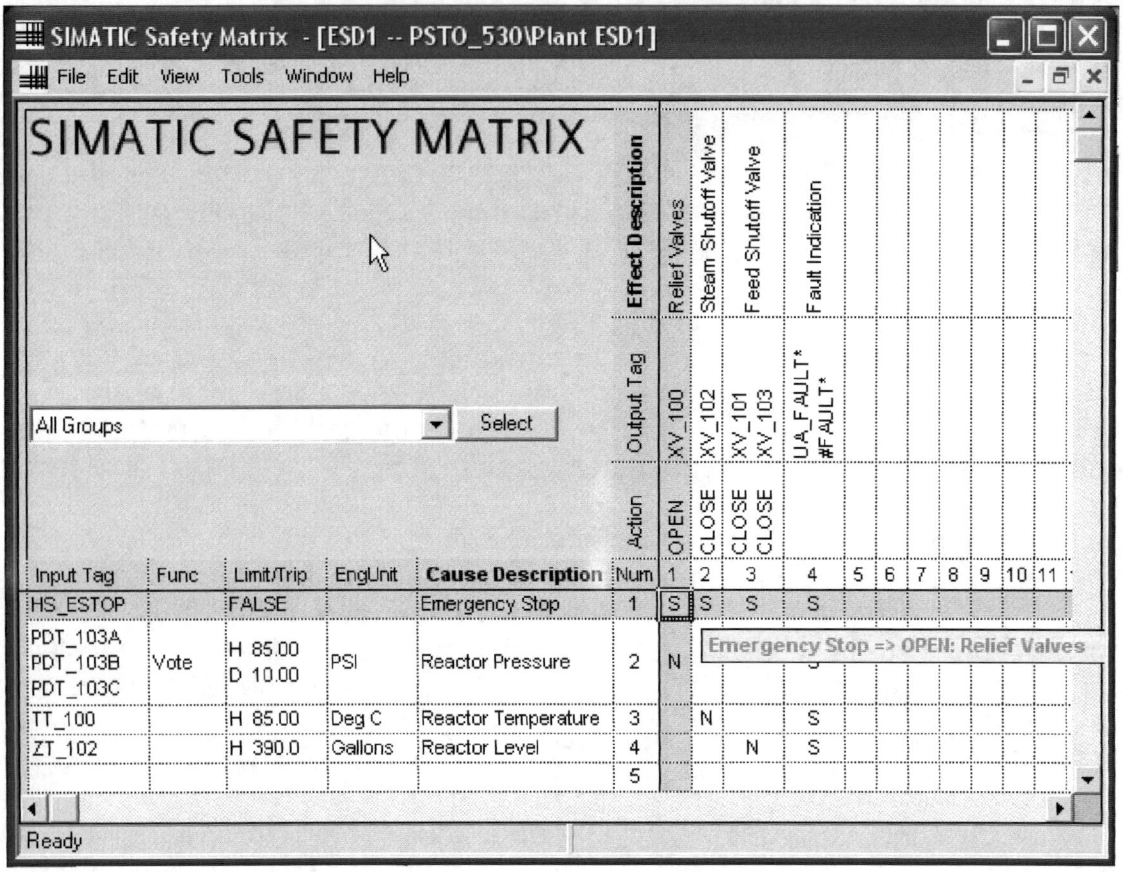

Ideally, there should no longer be the need for separate documents, databases, configurations, interpretations and translations for engineering, operations, maintenance and compliance verification - one document provides a clear, concise, maintainable and auditable reference.

Modern Safety Life-cycle tool has made achieving this ideal or model, a possibility. An example : SIMATIC Safety Matrix is an Engineering tool when used with SIMATIC S7-

Enhanced Safety from Advanced Simplicity

400FH addresses best part of the Safety Life-cycle issues mentioned earlier. The Siemens Fail-safe Safety Matrix programming package significantly simplifies Engineering and aids Change Management from Requirements (SRS) through the phases of Realization, Operation and Maintenance. The distinctiveness of SIMATIC Safety Matrix is that it makes programming **transparent, instantly recognizable, unmistakable** by means of **TUV tested and certified** automatic compilation into **Fail-safe** Continuous Function Chart (CFC).

The "by-design" Fail-safe Continuous Function Chart employs TÜV-certified function blocks from the S7 F (Fail-safe) System library. For advanced user there is the flexibility to directly use Continuous Function Chart, exclusive of Safety Matrix. It can be debated that not all logics and interlocks can be denoted by a Cause and Effect (C&E) Matrix. My recommendation is to always use the simplest form of representation such as a table or C&E Matrices to **aid comprehension** by the **various levels of people** in an organization **who utilize the information** through the different phases of the Life-cycle.

The SIEMENS SIMATIC Safety Matrix does provide another big enhancement - it permits operation and real-time monitoring of values of the safety application during runtime. The signal status is displayed with colors online in the Cause & Effect matrix. The online operation is possible from either the engineering station (monitor mode) or the operator HMI station (Safety Matrix Viewer).

Enhanced Safety from Advanced Simplicity

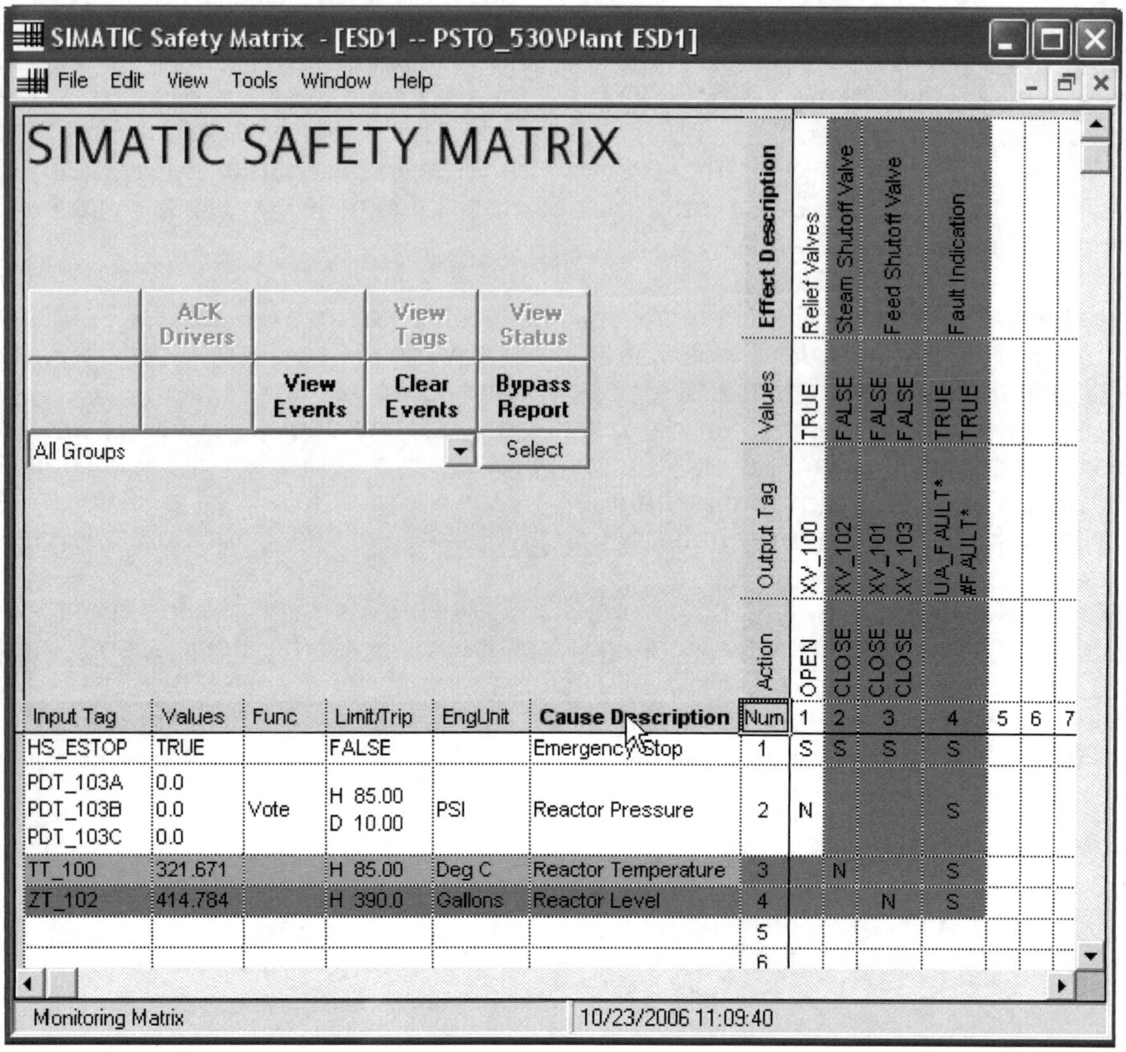

Safety is enhanced by providing a tested and certified tool for configuring the common SIS parameters such as voting, time delays, alarms, trips, bypass and override functions. The use of this built-in feature, in my assessment, greatly reduces the potential for non- compliance to standards and error introduction as a result of initial coding as well as subsequent modification.

To me the central benefit of simplifying is **Enhanced Overall Safety**. The related benefits that could be just as important are:

1) Document simplification – makes it easier to maintain documents current and "**updated**" – this will reinforce plant safety and compliance to Standards.

2) Significant reduction in engineering, testing, commissioning and training needs - **reduced overheads** (cost/time), resulting in faster project realization.

Enhanced Safety from Advanced Simplicity

Process Safety **Standards** IEC 61508 and IEC 61511, call attention to the significance of minimizing the **propagation of elemental failure** into system failure through correct "Safety Life-cycle" path - analysis, design, verification, operation, inspection and maintenance practices.

In the context of the modern-day globalized workplace and global workforce, the **different phases** of the "safety life-cycle" are performed by suitably trained personnel or specialists carrying out individual activities concurrently or at **different times**, often in **different time-zones**.

Potential for misrepresentations, misinterpretations due to language, cultural differences and "systematic" error introduction in the course of regular life-cycle activities, exists. Consequence of these can be lethal. Because Safety Matrix is a visual representation of causes, effects and interlocks it is easy to configure and "see" the system logic during coding, testing, operation, maintenance, upgrades or modifications. Furthermore, it self-documents.

Simplification in Safety Life-cycle management takes away the obscurity and the fear that mistakes would have crept in during the routine safety life-cycle activities.

"**Simplicity is prerequisite for reliability**. " – Edsger Dijkstra

....and so it is, for Safety.

Enhanced Safety from Advanced Simplicity

Session Selection:
"Enhanced Applications and Sharing of Lessons Learned"
Title: Enhanced Safety from Advanced Simplicity
E-mail Address for Permissions: stanleyjoseph8@gmail.com
Preferred Presentation Format: Oral and Print
Author and Presenting Author: Stanley Joseph
Email: stanley.joseph@siemens.com

Enhancing Safety Knowledge in Students: The IIT Gandhinagar Experience

Raj Srinivasan
Indian Institute of Technology Gandhinagar
Vishwakarma Government Engineering College Complex
Chandkheda, Ahmedabad
India - 382424
raj@iitgn.ac.in

Heroes, Villains & Victims

"Federal investigators say quick thinking by the crew of US Airways Flight 1549, coupled with the plane's safety equipment, contributed to the successful river landing of the crippled airliner" Wired, May 2010

Heroes, Villains & Victims

"Airline pilots were once the heroes of the skies. Today, in the quest for safety, airplanes are meant to largely fly themselves. Which is why the 2009 crash of Air France Flight 447, which killed 228 people, remains so perplexing and significant." Vanity Fair, Oct 2014

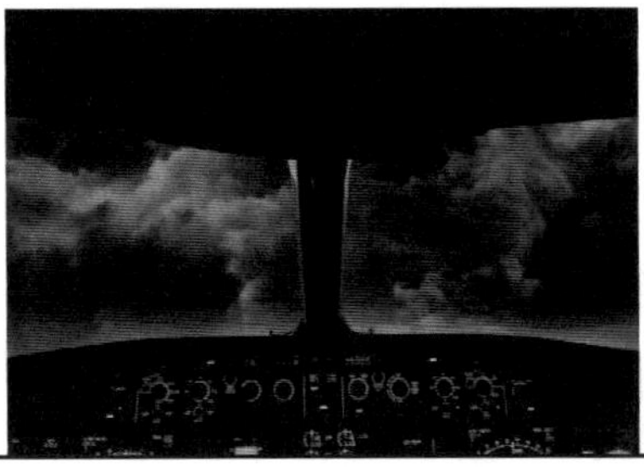

Heroes, Villains & Victims

"It seems that we are locked into a spiral in which poor human performance begets automation, which worsens human performance, which begets increasing automation …In the aftermath of the accident, the pitot tubes were replaced on several Airbus models …independent safety review .. highlighted the arrogance of some of the company's pilots … a number of experts called for angle-of-attack indicators in airliners, while others urged a new emphasis on high-altitude-stall training, upset recoveries, unusual attitudes, flying in Alternate Law, and basic aeronautical common sense."

Heroes, Villains & Victims

"..none of it will make much difference. At a time when accidents are extremely rare, each one becomes a one-off event, unlikely to be repeated in detail. Next time it will be some other airline, some other culture, and some other failure—but it will almost certainly involve automation and will perplex us when it occurs. Over time the automation will expand to handle in-flight failures and emergencies, and as the safety record improves, pilots will gradually be squeezed from the cockpit altogether. The dynamic has become inevitable. There will still be accidents, but at some point we will have only the machines to blame."

What is the Role of Humans?

- Human Factors Engineering
  - "concerned with the understanding of interactions among humans and other elements of a system, …to optimize human well-being and overall system performance"
- Self-driving cars, robots for assembly line automation, lights-off operation in chemical plants…

- As society embraces complexity, what is the role of humans?

The Wheel of Fortune

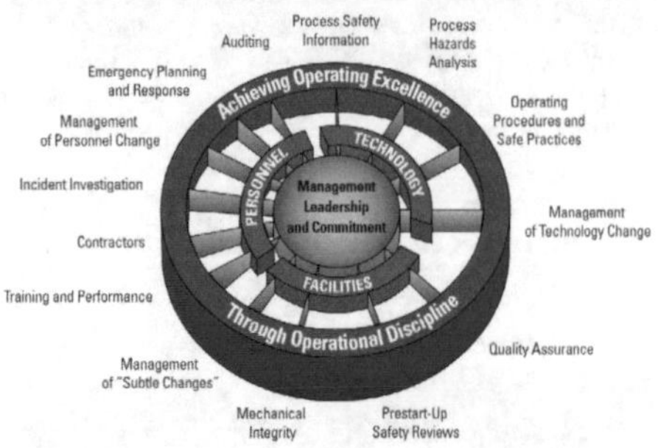

IIT Gandhinagar

- **Started operations in 2008**
- **Currently ~1000 students ~100 faculty**
- **Initial target: 2,500 students and 250 faculty**
- **Academic Programmes**
  - B.Tech and M.Tech in Engineering
  - M.Sc. in Sciences (including **Cognitive Science**)
  - M.A. in **Society and Culture** (from 2014)
  - Ph.D in all areas of Humanities, Sciences and Engineering
- **Permanent campus under construction**

Safety Centre @ IITGN

- Mission
  - » Discover
  - » Teach
  - » Preach (outreach)
  - » Practice
- Domains
  - » Process safety
  - » Seismic safety
  - » Road safety
  - ➢ Fire engineering
  - ➢ Cognitive engineering

Safety Centre Faculty

Practice

- Safety Workshop for all staff
  - » Training on Hazard identification, risk assessment
- Emergency response & fire fighting
- Incident investigation

Preach

About 1400 participants to date

International Conference on Safety, Dec 2 -6 2014 ....Have we made progress since Bhopal?

Teach

- Foundation program
- Courses in the curriculum
  - » Fundamentals of Process Safety
  - » Introduction to Fire Engineering
  - » Structures under fire
  - » Compulsory safety course for all graduate students <starting 2015>
- Industrial visits
- AIChE SAChE modules

UL Engineering Challenges

- 2010: National Fire Database
- 2011: Innovative PV Applications
- 2012: Electric Vehicle Challenge
- 2013: Safer Water Challenge
- 2014: Electrical Safety

PROCESS SAFETY RESEARCH

Motivation

- 30 years since Bhopal
- 20+ years since OSHA PSM standards, Seveso II Directives, etc
- Major accidents continue

Buncefield, UK 2005 | Jaipur, India 2009 | West, Texas USA 2013

- Need to analyze process safety & accidents
  - » Are there any trends?

How do we know what we know?

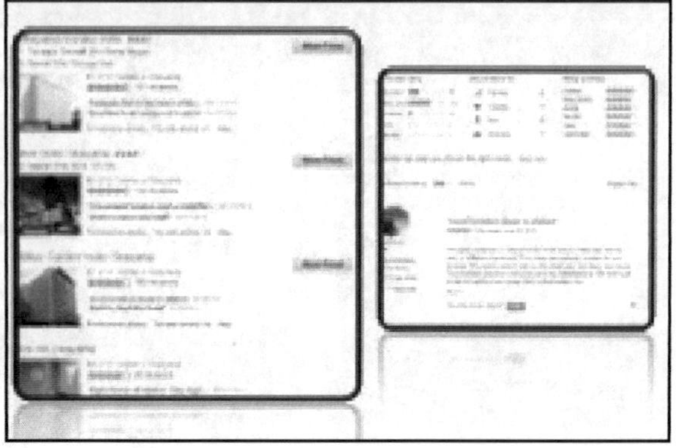

Approach

- Text mining is used to analyze various safety related textual sources
- Sources
  - » Corporate sustainability reports
  - » Google books
  - » Scopus
  - » Accident investigation reports published by CSB, EPA (in USA) and OISD (in India)
- Software: RapidMiner, Microsoft Excel

Corporate Sustainability Reports

- 112 Companies from Forbes 2000 list
  - » Oil & Gas
    - BP, Chevron, Conoco-Philips, Petrobras,...
  - » Chemicals
    - Air products, BASF, Celanese...
  - » Pharmaceuticals
    - Abbott, Amgen, Pfizer, Roche...
  - » Consumer products
    - Avon, Colgate, Danone, Nestle, Unilever...

Unilever

Source: http://www.unilever.com/sustainable-living-2014/

Top 10 Sustainability Related Phrases

Rank	Chemicals	Consumer Products	Oil/Petrochemicals	Pharmaceuticals
1	Raw material	Raw material	Human rights	Health and safety
2	GHG emission	Human rights	GHG emission	Human rights
3	Climate change	Health and safety	Climate change	Access to medicine
4	Health and safety	Environmental impact	Energy efficiency	GHG emission
5	Human rights	GHG emission	Health and safety	Climate change
6	Responsible Care	CO2 emission	Oil spill*	Environmental impact
7	Process safety	Water use	Environmental impact	CO2 emission
8	Hazard management	Life cycle	Local community*	Hazard management
9	CO2 emission	Energy consumption	Hazard management	Hazardous waste
10	Energy consumption	Product safety	CO2 emission	Raw material

Several topics are consistently of importance in all 4 sectors

Top 10 Sustainability Related Phrases

Rank	Chemicals	Consumer Products	Oil/Petrochemicals	Pharmaceuticals
1	Raw material	Raw material	Human rights	Health and safety
2	GHG emission	Human rights	GHG emission	Human rights
3	Climate change	Health and safety	Climate change	Access to medicine
4	Health and safety	Environmental impact	Energy efficiency	GHG emission
5	Human rights	GHG emission	Health and safety	Climate change
6	Responsible Care	CO2 emission	Oil spill	Environmental impact
7	Process safety	Water use	Environmental impact	CO2 emission
8	Hazard management	Life cycle	Local community	Hazard management
9	CO2 emission	Energy consumption	Hazard management	Hazardous waste
10	Energy consumption	Product safety	CO2 emission	Raw material

*Globally unique
*Unique within top 10

Hierarchy of Sustainability Concepts

- Structure to organize various sustainability concepts
  - Constructed by looking at the top sustainability-related terms
  - Organized based on general understanding of sustainability

Ranking of Sustainability Concepts

Sustainability Issues	Chem	Consumer	Oil	Pharma
Health and safety	1	1	1	1
Human rights	3	2	2	2
Reducing GHG	2	3	4	3
Conserving energy	4	4	3	4
Community investment	5	7	5	5
Reducing waste	7	5	10	6
Renewable energy	6	8	6	9
Conserving water	9	6	8	7
Employee's benefits	8	9	9	8
Reducing VOC	10	14	13	11
Conserving natural resource	11	11	14	12
Biodiversity preservation	13	12	12	13
Recycling	12	10	15	14
Green chemistry	14	13	*17.5*	10
Local supplier/content	15	15	11	16
Preventing oil spill	*18*	*17.5*	7	*17.5*
Conserving solvent	16	*17.5*	*17.5*	15
Reducing natural gas	17	16	16	*17.5*

Ranking of Sustainability Concepts

Sustainability Issues	Chem	Consumer	Oil	Pharma	Average	Min	Max	Range
Health and safety	1	1	1	1	1	1	1	0
Human rights	3	2	2	2	2.25	2	3	1
Reducing GHG	2	3	4	3	3	2	4	2
Conserving energy	4	4	3	4	3.75	3	4	1
Community investment	5	7	5	5	5.5	5	7	2
Reducing waste	7	5	10	6	7	5	10	5
Renewable energy	6	8	6	9	7.25	6	9	3
Conserving water	9	6	8	7	7.5	6	9	3
Employee's benefits	8	9	9	8	8.5	8	9	1
Reducing VOC	10	14	13	11	12	10	14	4
Conserving natural resource	11	11	14	12	12	11	14	3
Biodiversity preservation	13	12	12	13	12.5	12	13	1
Recycling	12	10	15	14	12.75	10	15	5
Green chemistry	14	13	*17.5*	10	13.625	10	17.5	7.5
Local supplier/content	15	15	11	16	14.25	11	16	5
Preventing oil spill	*18*	*17.5*	7	*17.5*	15	7	18	11
Conserving solvent	16	*17.5*	*17.5*	15	16.5	15	17.5	2.5
Reducing natural gas	17	16	16	*17.5*	16.625	16	17.5	1.5

Sustainability Hierarchy

		Conclusion			
		Level 1	Level 2	Level 3	Level 4
Premise	Level 1	N.A.	72%	30%	0%
	Level 2	92%	N.A.	67%	5%
	Level 3	89%	86%	N.A.	15%
	Level 4	85%	91%	85%	N.A.

- Level 4: Gender equality
- Level 3: Customer privacy, indigenous people, material reduction
- Level 2: Save energy, reduce waste, forced labor, child labor, biodiversity protection, customer safety, equal opportunity
- Level 1: Emission reduction, occupational safety, renewable energy, spillage, local community, water recycle
  - Companies move to higher level once lower levels satisfied

Sustainability Hierarchy

- Social issues at the top
- Environmental issues predominant at the bottom

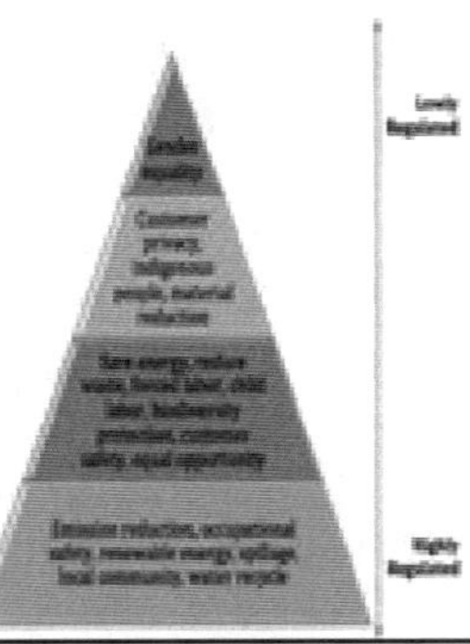

Process Safety in Books

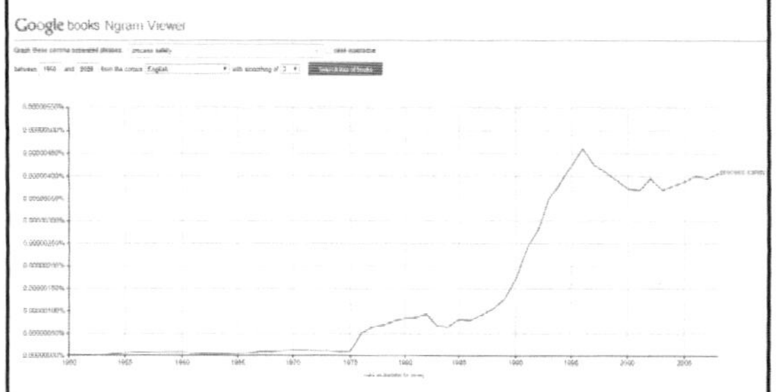

Process Safety in Books

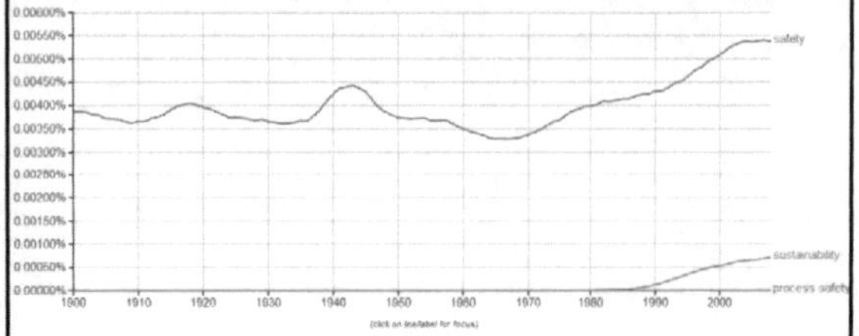

Process Safety in Scopus

Process Safety in Scopus

Trade Publications	Journals
Accident-prevention	Accident-prevention
Process-control	Risk-assessment
Risk-assessment	Process-safety-management
Risk-management	Human
Standards	Safety-engineering
Process-safety-management	Risk-management
Process-engineering	Hazards
Regulatory-compliance	Process-control
Hazards	Accidents
Laws-and-legislation	Mathematical-models

Process Safety in Scopus

- Common themes
  - » Accident prevention; Process safety management
  - » Risk assessment; Risk management; Hazards
- Themes emphasized in trade publications
  - » Standards, regulatory compliance, benchmarking
  - » Maintenance, automation, personnel, personnel-training, costs
- Themes emphasized in journals
  - » Safety engineering; design; risk analysis
  - » Random processes; mathematical models; computer simulation' optimization

Analysis of accident reports

- Corpus
  - » 'Incident Analysis' & 'Recommendations' section from 34 CSB reports, 8 EPA reports
- Size – Max / Min / Avg
  - » CSB: 341 / 33 / 94 pages
  - » EPA: 114 / 50 / 77 pages
- Top 200 terms that are over-expressed in these reports identified
  - » Relative TF-IDF analysis
  - » Safety words

Making Sense: Words to Concepts

The main reason of the LPG leakage [MECHANICAL FAILURE] (of loading arm content only) was the failure of the said interlock [SAFETY PROVISION] of the coupling. It appears that LPG [CHEMICAL] gushed out with high velocity under the operating pressure and the same resulted in recoil of the loading arm and/or the coupling. Due to this recoil of the loading arm and/or the coupling, it might have collided with the adjoining structure resulting in a source of ignition [INITIATION] creating the flash fire [OUTCOME] with the leaking LPG pool creating enough vapor to produce a combustible mixture in the open.

Red = safety terms
= concepts

Concept analysis

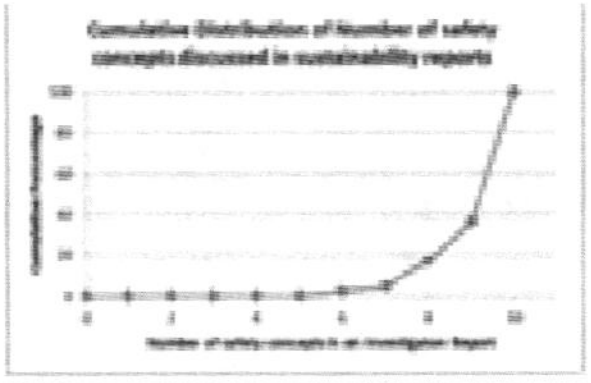

Total Number of Key Concepts identified	11
Average Number of Concepts	9.39
Maximum number of concepts	10
Minimum number of concepts	6
Standard Deviation of topics	0.971

Concept Analysis

Safety Concept	Reports
Safety Provision	98%
Preventive Measure	98%
Hazard	98%
Chemical	98%
Outcome	95%
Management	95%
Equipment	95%
Mechanical failure	83%
Initiation	81%
Emergency Planning	76%

Validation of Concept Tree

Validation using 20 OISD reports: 11 / 2 / 5 pages

CSB and EPA reports

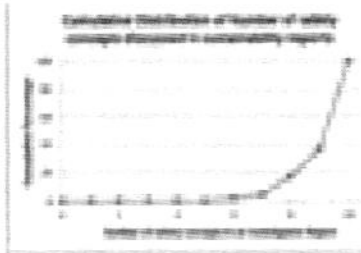

Total Number of Key Concepts	11
Average number of concepts/report	9.39
Maximum number of concepts	10
Minimum number of concepts	6
Standard Deviation	0.971

OISD reports

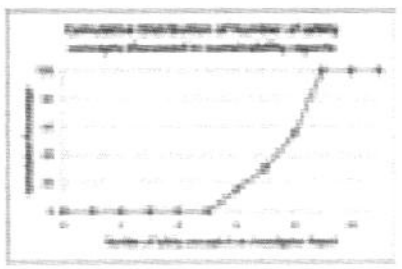

Total Number of Key Concepts	11
Average number of concepts	8
Maximum number of concepts	9
Minimum number of concepts	6
Standard Deviation topics	1.12

OISD Concept Analysis

Safety Concept	OISD Reports	CSB / EPA Reports
Equipment	95%	95%
Safety Provision	90%	98%
Chemical	90%	98%
Outcome	85%	95%
Management	85%	95%
Hazard	65%	98%
Initiation	60%	81%
Mechanical failure	45%	83%
Emergency planning	35%	76%
Preventive Measure	20%	98%

The Paradox

- " People don't release smoke, grind gears, or have pieces fall off; but their equivalents – fatigue, error, injury, and ill health – do result in failure and breakdown."
- " We hire people to watch the equipment in our factories, nuclear power plants, and airplanes, but we don't have the equipment watch the person to ensure that he or she is awake and alert. We have the weak link of the chain watching the strong link, but not the strong watching the weak."

Addison-Wesley (Feb 1994)

Eye Tracking

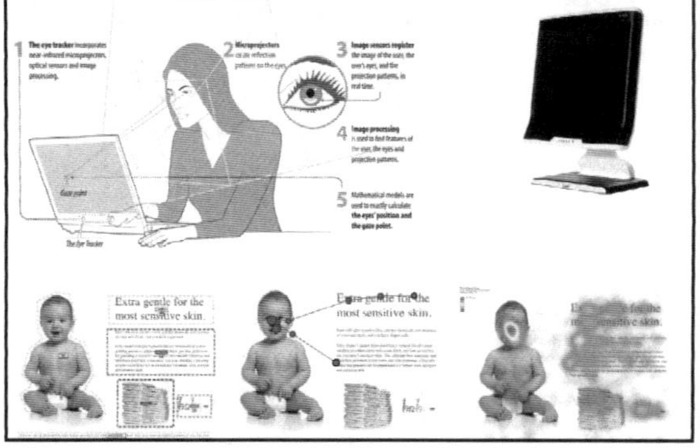

Eye Tracking Examples

Conscious Driving **Subconscious Driving**

Eye Tracking Examples

Familiar route **Unfamiliar route**

Other Applications

- Medicine
  - » Surgeon skill assessment
  - » Improvement of training program
- Aviation
  - » Situation awareness
  - » Attention tunnelling
- Monitoring cognitive activity of drivers

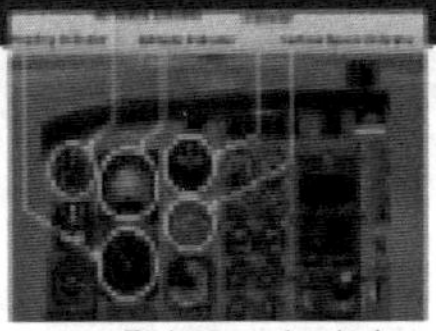

Key instruments in cruise phase

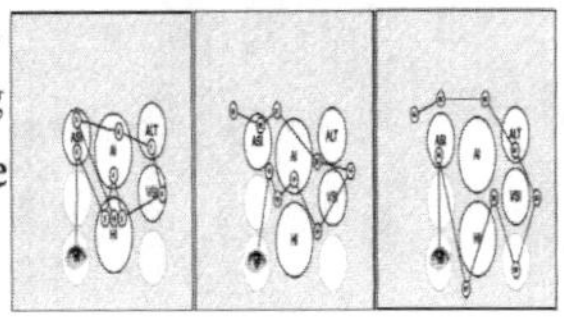

(a) Attention Focusing (b) Attention Blurring (c) Misplaced Attention
(Kilingaru et al. Intelligent and Fuzzy systems 2013)

A Simple Process

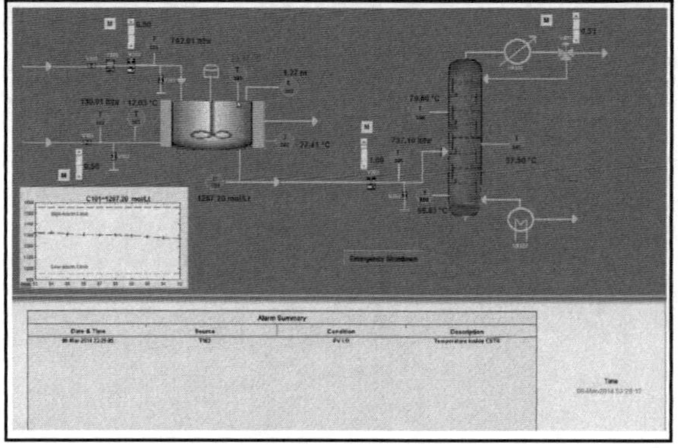

Video of Eye Tracking

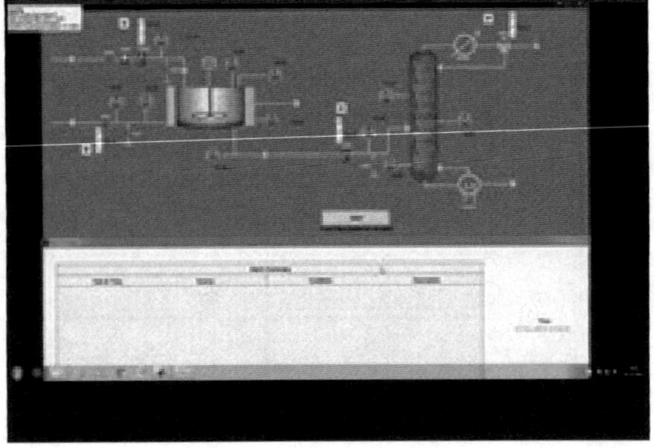

Typical Operator Behavior

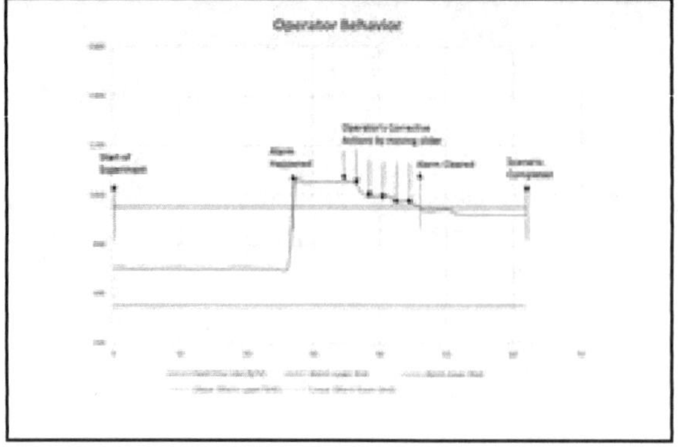

Typical Operator Behavior

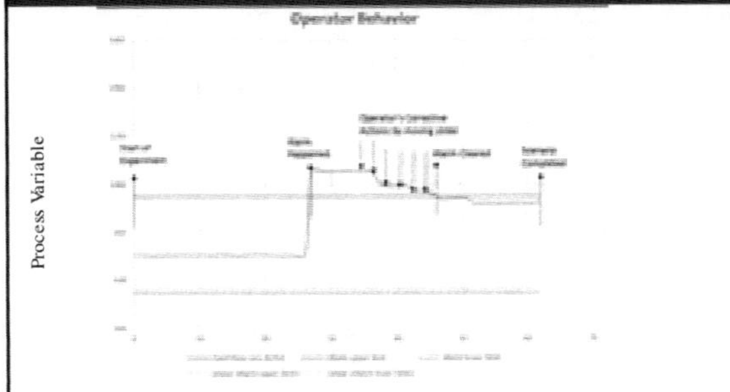

- Av. time delay to detect tag after alarm: 1.4 s
- Av. time spent during first gaze at alarm tag: 0.63 s / Pupil dilation: 0.33 mm
- Av. delay before first click on MV: 6.83 sec

First Visual Action after alarm

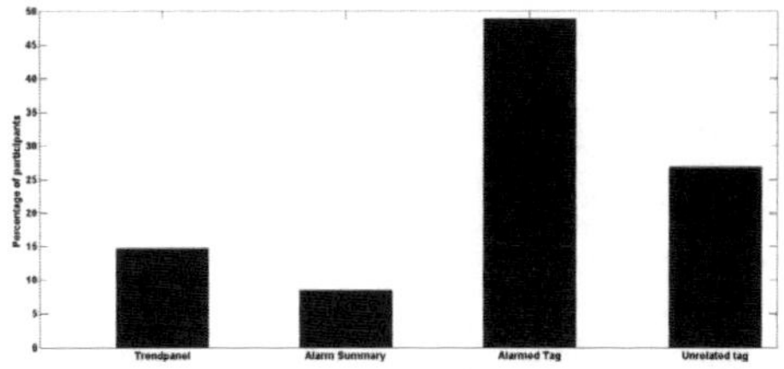

Gaze Distribution

Before Alarm

After Alarm

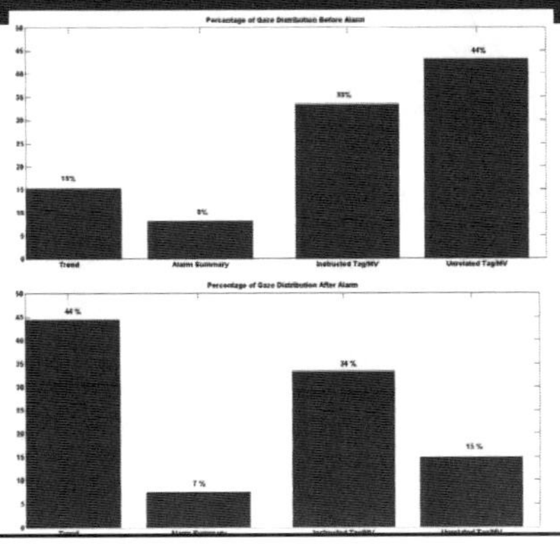

Gaze Time Distribution (before alarm)

Success cases

Failure cases

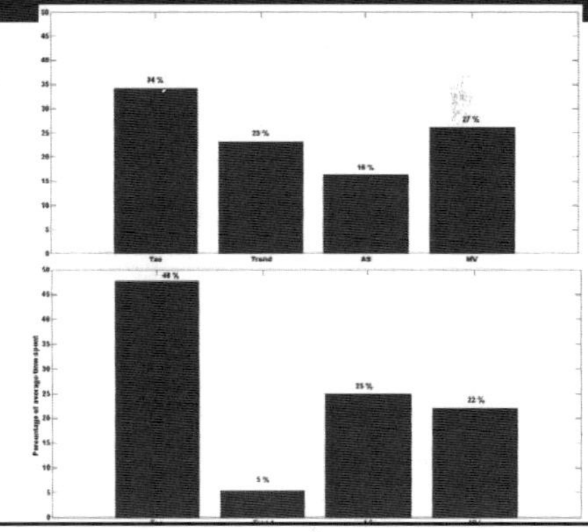

A "Clear-thinking" Operator

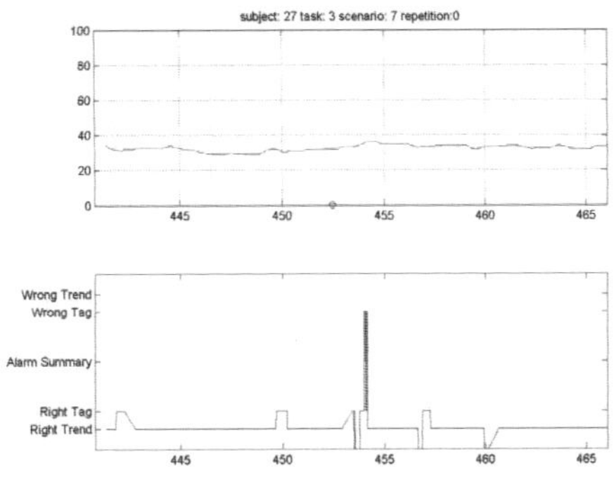

A "muddled" Operator

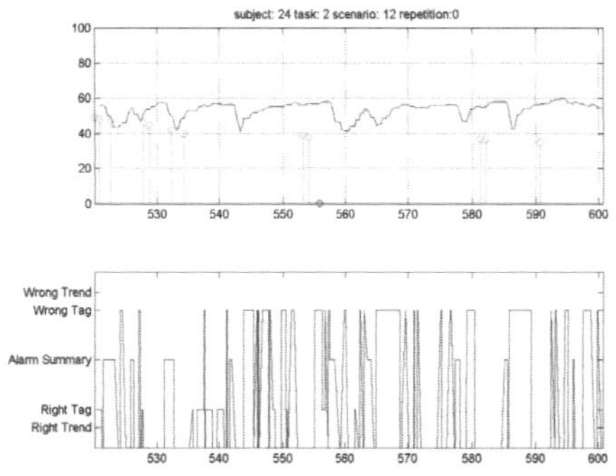

Research Goals

- Human behavior
  - » Goal: From open-loop towards closed-loop
  - » First step: Sensing & sense-making
- Current work
  - » Index of operator's attention level
- Future work
  - » Other measurements - Stress level

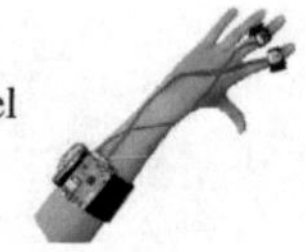

In Closing…

- Safety Centre Mission
  - » Discover
  - » Teach
  - » Preach
  - » Practice
- Beginning of journey
- Invite companies to join hands with us

Enhancing Safety Knowledge in Students:
The IIT Gandhinagar Experience

Thank you !

Raj Srinivasan

Indian Institute of Technology Gandhinagar

Vishwakarma Government Engineering College Complex

Chandkheda, Ahmedabad

India - 382424

raj@iitgn.ac.in

OPERATIONS SAFETY HUDDLE

V V R Narasimham
Executive Director – HSE Corporate

Hindustan Petroleum Corporation Ltd

1

FIGURE THIS OUT...

IN MOST CASES OF OPERATIONS:

- **THE FINAL LAST ACT LEADING TO AN INCIDENT WILL BE BY A** LOWEST LEVEL PLANT TECHNICIAN
  - **Could be Mistake...**
  - **Could be Negligence...**
- **But** management is answerable

2

IN THAT CASE...

- **What is being done to influence his Safety Culture?**

3

What is in his mind ?

- **How much does he think of Safety?**
- **Is he convinced that his immediate Supervisors are committed to safety?**
- **Does he realise that he could be repeating mistakes?**
- **Is he trapped in "theory" v/s "practical" syndrome?**
- **Does he think that management is serious about Safety or only Shows off**

4

RECOGNISE ...

- **These thoughts and feelings of the frontline – They matter.**
- **This will influence their actions.**
- **The "Reality" of management's genuine concern on Safety must match with the "perception" of the rank and file.**

5

Management Questions...

- **Are his supervisors regularly talking about Safety?**
- **Are they providing direction & guidance?**
- **How do his immediate supervisors conduct themselves in Plant operations?**
- **When was the last time a senior person spoke to them about safety?**
- **Is there space for open discussions?**
- **Is there a system of regular knowledge inputs ?**
- **Are we sharing Incident findings?**

6

Safety Leadership...

- Irrigate the minds with Safety principles.
- Impart knowledge
- Bring openness & transparency
- Engage in Safety discussions
- Maintain momentum
- Demonstrate Safety Commitment

7

OPERATIONS SAFETY HUDDLE

- To enhance engagement in Safety
- To embed Safety in work routine.
- Increase knowledge about the Plant they work in
- Generate interest in Safe working
- Generate positive perception about management's concern about safety.

8

OPERATIONS SAFETY HUDDLE

- **Every shift starts with the Huddle.**
- **Compulsory 10 minutes interaction.**
- **Conducted by Shift incharge**
- **Separate in each Plant**
- **Content structured, part pre-planned**
- **Implemented in January, 2011. Rev-1 in 2013.**

9

OPERATIONS SAFETY HUDDLE

- **Senior management Visits.**
- **Cross Visits by Operations Division Heads**
- **Safety Shift Officer Visit.**
- **Quizzes**
- **Gifts**

10

OPERATIONS SAFETY HUDDLE

Initial reactions :

- Why is it required
- Work will get delayed
- Many supervisors cannot talk well
- Cannot conduct on start up, shutdown days
- Mutual blame discussions will be there
- One more system
- Let us be practical .....

11

OPERATIONS SAFETY HUDDLE

- Process in PDCA model.
- Standard template.
- Info bank of incidents on Intranet portal.
- Time Table for imparting Process Safety Information.
- Methodology of imparting information specified.
- Review reports

12

OPERATIONS SAFETY HUDDLE

Experience

- **Only 2 out of 5 Divisions initiated**
- **Others took two weeks to "start"**
- **After 3 months, 60-70% satisfaction.**
- **Every incident of mistake must be used to drill down safety systems.**
- **95% satisfaction in 1 year.**
- **Quality satisfaction in about 70%.**

13

OPERATIONS SAFETY HUDDLE

Good media to impart Process safety Information

- **OSHA-PSM requires dissemination of Process Safety Information to the employees.**
- **Information : Hazardous chemicals, technology, Equipment**
- **Methodology and Time Table of imparting made and incorporated.**

14

OPERATIONS SAFETY HUDDLE

Benefits

- Knowledge levels of Plant Operators increased.
- Understanding of Plant features improved- especially trips and interlocks, Relief System.
- Compliance of "Planned Arrangement" termed DRJ-Calendar improved.
- Enhanced perception of management concern on Safety.

15

OPERATIONS SAFETY HUDDLE

Benefits

- Openness on mistakes & reduction in mistakes.
- Discussion on what can go wrong before doing any job.
- JHA compliance improved in maintenance works.
- More buy-in into Management's initiatives.

16

OPERATIONS SAFETY HUDDLE

17

OPERATIONS SAFETY HUDDLE

Good Tool for :

- **Improving frontline engagement**
- **Improving Plant Knowledge**
- **Continual Safety Enhancement**
- **Help in OSHA-PSM Requirements**

18

THANKS

19

CCPS An AIChE Technology Alliance Center for Chemical Process Safety

VISION 20/20

Vision 20/20
A Call to Action

Shakeel Kadri
Director, Global Process Safety & Risk Management
Air Products and Chemicals, Inc
and
CCPS Executive Director Designate

15 December 2014
CCPS Global Summit on Process Safety, Mumbai, India
(Adapted from the 10th Global Congress on Process Safety presentation)

CCPS An AIChE Technology Alliance Center for Chemical Process Safety

Why Create a Vision?

- Our common goal - Reduce incidents
- Why the Vision?
  - We want to reach the goal faster
  - Leverage our collective strength
  - Help CCPS and other organizations identify projects/initiatives to help us reach the goal
  - Make a step change in management of process safety

It's time to leverage our resources, knowledge and skills to all strive for a common goal of great process safety performance. CCPS's Vision 20/20 describes that vision.

CCPS — An AIChE Technology Alliance — Center for Chemical Process Safety

Vision 20/20
The CCPS Project

- Envisions the characteristics of companies with great process safety performance in the future, 2020
- Committee members
  - **Cheryl Grounds - Chair**
  - **Dave Jones**
  - **Jack McCavit – CCPS**
  - **Jeff Fox**
  - **Joe Allaben**
  - **Karen Tancredi**
  - **Louisa Nara**
  - **Mike Broadribb**
  - **Pete Lodal**
  - **Samantha Scruggs**
  - **Scott Berger**
  - **Steve Arendt**
  - **Todd Aukerman**
  - **Walt Frank**

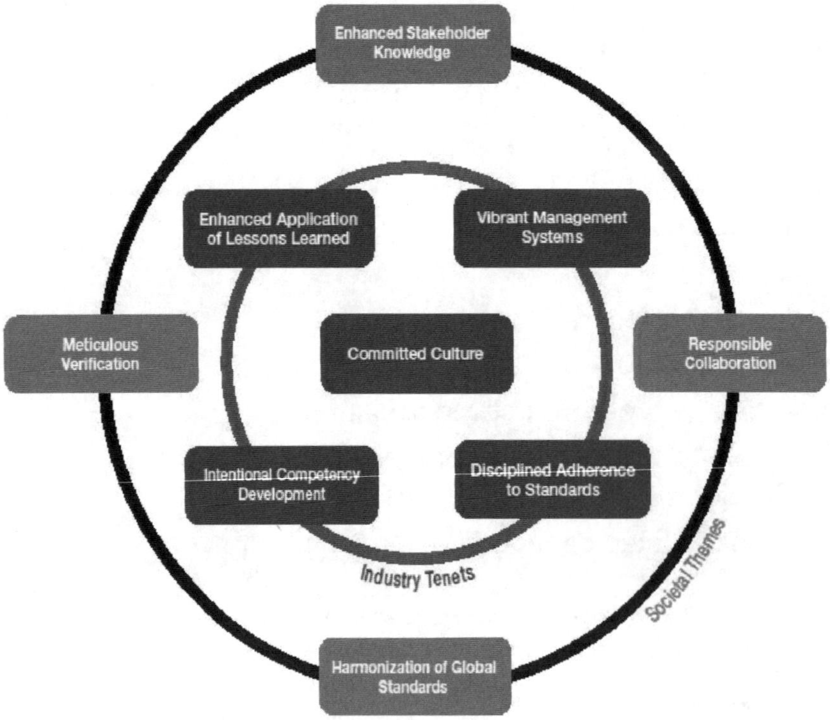

Envisioning the Future...

Company Tenets

1. Committed Culture,
2. Vibrant Management Systems,
3. Disciplined Adherence to Standards,
4. Intentional Competency Development, and
5. Enhanced Application and Sharing of Lessons Learned.

Societal Themes

1. Enhanced Stakeholder Knowledge,
2. Responsible Collaboration,
3. Harmonization of Global Standards, and
4. Meticulous Verification

... achieving great process safety performance

Committed Culture

- Felt leadership from senior executives through plant personnel
- Leadership instills disciplined adherence to the policies and procedures
- Strong sense of vulnerability
- What can you do?
  - ☑ Perform all tasks, correctly, every time.
  - ☑ Support your leadership in demonstrating a commitment to process safety.
  - ☑ Recognize that "it could happen here..."

Vibrant Management Systems

- Structure that clearly defines expectations for all systems that impact process safety performance
- Promotes principles of inherently safer design
- Promotes principles of *Guidelines for Risk Based Process Safety* including fit-for-purpose policies and procedures
- What can you do"
  - ☑ Do your part as a member of the team - use the system!
  - ☑ Share your learnings and suggested improvements to the system.

Disciplined Adherence to Standards

- Company and/or industry standards rigorously followed
- System to ensure that existing equipment meets company minimum expectations
- Risk decision making is used when applicable standards do not exist
- What can you do?
  - ☑ Ensure a system exists to effectively use industry and company standards.
  - ☑ Make sure that the system applies to existing equipment.
  - ☑ As a leader, assure standards are followed and respond accordingly.

Intentional Competency Development

- Ensures that employees who impact process safety performance are meeting the technical and cultural requirements of their job
- It is given that companies provide sufficient numbers of employees

- What can you do?
  - ☑ Develop a plan to fill your competency gaps.
  - ☑ When you're outside your area of competency, contact an expert that can help.
  - ☑ As a leader, build and recognize team member competence.

Enhanced Application of Lessons Learned

- Thirst for learning from
  - Incidents
  - Near misses
  - Benchmarking
  - Jobs done well
- Changes are implemented based on the lessons Learned

- What can you do?
  - ☑ Use a significant incident or near miss as an opportunity to encourage and embed learnings.
  - ☑ Take a personal action, or make a change, as a result of the learning. Sharing only makes you feel good, learning makes a difference.

Envisioning the Future...

Company Tenets	Societal Themes

Company Tenets

1. Committed Culture,
2. Vibrant Management Systems,
3. Disciplined Adherence to Standards,
4. Intentional Competency Development, and
5. Enhanced Application and Sharing of Lessons Learned.

Societal Themes

1. **Enhanced Stakeholder Knowledge,**
2. **Responsible Collaboration,**
3. **Harmonization of Global Standards, and**
4. **Meticulous Verification**

... achieving great process safety performance

Enhanced Stakeholder Knowledge

- Risk literacy
  - Public
  - Government
  - Industry decision makers
- Process safety fundamentals
  - Chemical engineering students
  - Employees
- What can you do?

☑ Support local STEM education, both in high school and university.

☑ Have a conversation about risk with students and other stakeholders within your community.

☑ Continue to build knowledge throughout your career.

An AIChE Technology Alliance
CCPS
Center for Chemical Process Safety

Responsible Collaboration

- Technical, government, labor, community, scientific, academic, and industrial organizations work together
- Remove barriers to report incidents
- What can you do?

☑ Recognize the value that others bring to the table and commit to working together toward the common process safety goal.

☑ Be open-minded and focus on that common goal.

☑ Actively participate in or lead joint industry projects.

An AIChE Technology Alliance
CCPS
Center for Chemical Process Safety

Harmonization of Global Standards

- Standards can come from many sources
- Standards can be confusing and conflicting
- In the future, work jointly to streamline practices, eliminate redundancy, and cooperatively address emerging issues.
- What can you do?

☑ Recognize the diversity in standards, be open-minded, and work towards harmonization of standards.

☑ Promote harmonization through active participation in standard and practice writing groups.

CCPS An AIChE Technology Alliance
Center for Chemical Process Safety

Meticulous Verification

- Third parties including
  - Public
  - Non-Governmental Organizations
- Evaluate implementation of process safety programs
- What can you do?

☑ Identify opportunities where third-party verification could add value.
☑ Personally support external verification in your area when it occurs.

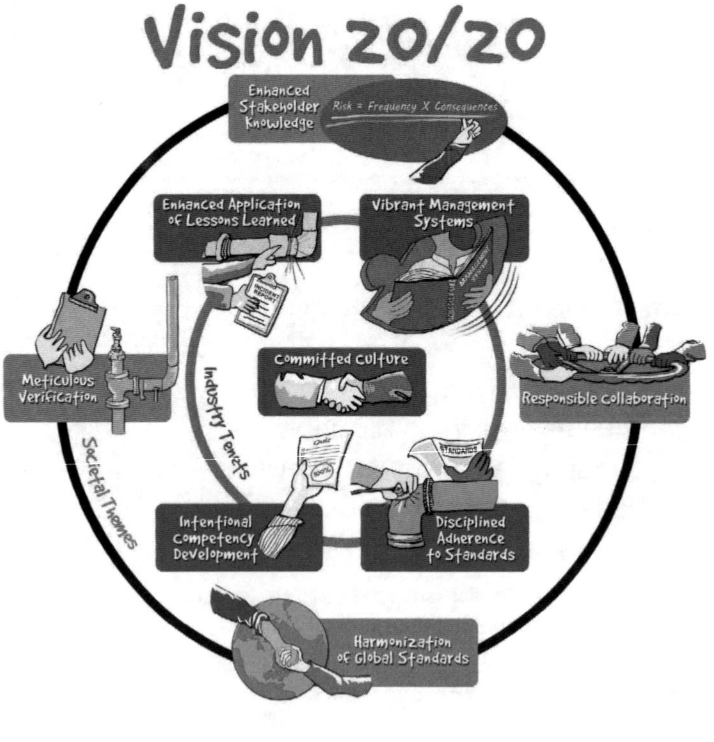

CCPS
An AIChE Technology Alliance
Center for Chemical Process Safety

A Call To Action

It's time to leverage our resources, knowledge and skills to all strive for a common goal of great process safety performance. CCPS's Vision 20/20 describes that vision.

- Understand and communicate the five tenets and themes
- Evaluate your contribution
  - within your company, across your discipline industry colleagues, and with your regulatory, academic, and local communities.
- Evaluate performance, seek collaboration and take action

CCPS
An AIChE Technology Alliance
Center for Chemical Process Safety

Vision 20/20 Tools - Current

- Brochure
- "A day in the life of"
- One page documents
- Booklets
- Presentations
  - Global Congress on Process Safety, New Orleans
  - IChemE Hazards Conference, Edinburg
  - Global Summit on Process Safety - Mumbai

Vision 20/20 Tools - Draft

- Assessment tool
  - Self assessment by representative team
  - Addresses each industry tenet
  - Intended to be
    - Easy to use
    - Differentiating
- Link to existing resources
  - CCPS texts
  - Worldwide industry resources

On the CCPS homepage

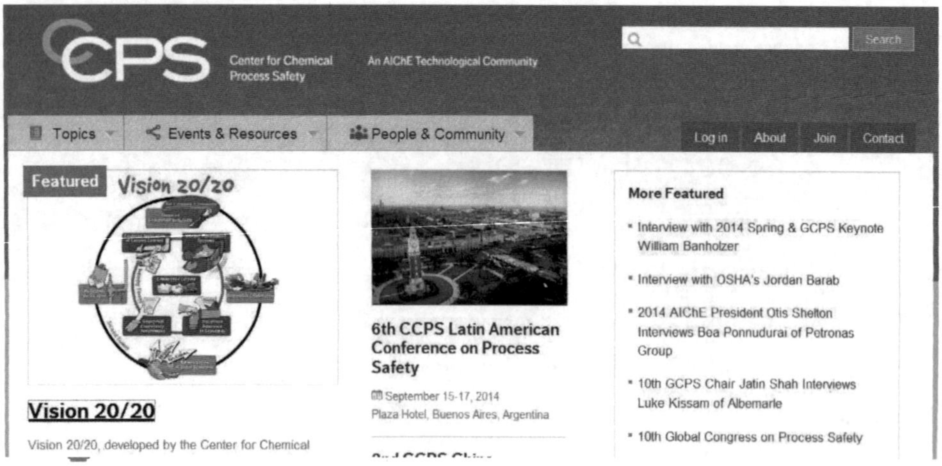

Thank you! Questions?

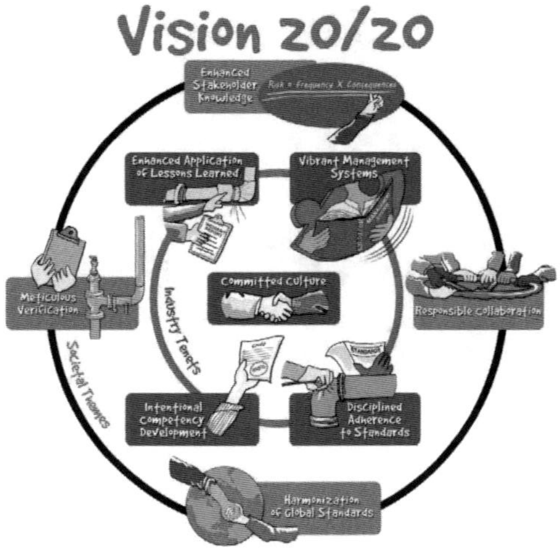

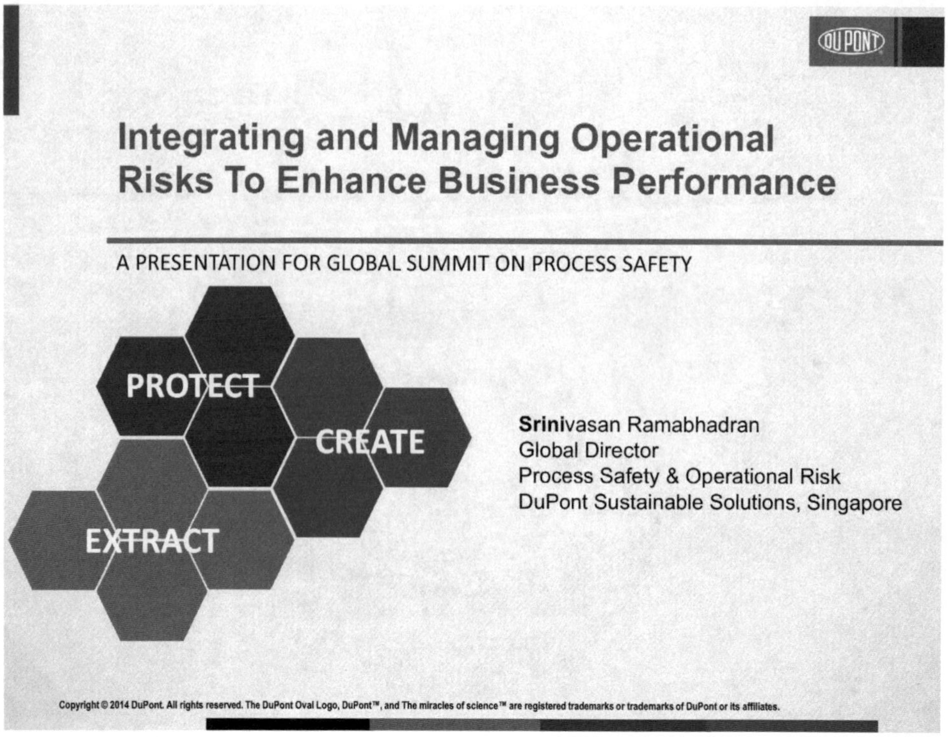
DUPONT
Integrating and Managing Operational Risks To Enhance Business Performance
A PRESENTATION FOR GLOBAL SUMMIT ON PROCESS SAFETY
PROTECT
CREATE
EXTRACT
Srinivasan Ramabhadran
Global Director
Process Safety & Operational Risk
DuPont Sustainable Solutions, Singapore
Copyright © 2014 DuPont. All rights reserved. The DuPont Oval Logo, DuPont™, and The miracles of science™ are registered trademarks or trademarks of DuPont or its affiliates.

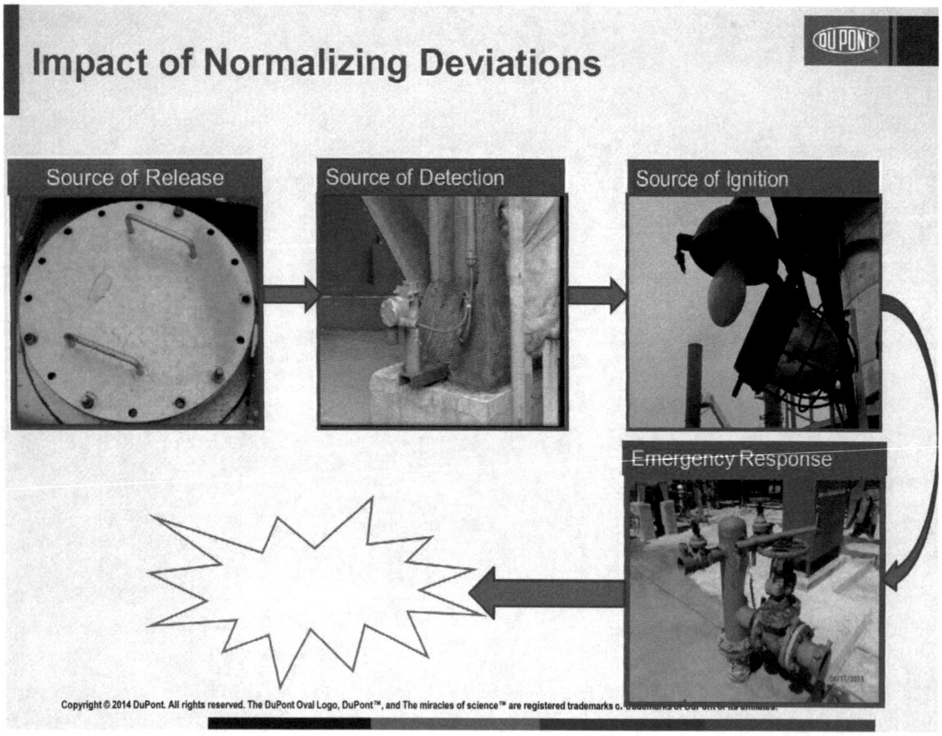
DUPONT
Impact of Normalizing Deviations
Source of Release
Source of Detection
Source of Ignition
Emergency Response
Copyright © 2014 DuPont. All rights reserved. The DuPont Oval Logo, DuPont™, and The miracles of science™ are registered trademarks

Key Drivers for Innovation in PSM

DU PONT

Top of mind issue for CEOs / Board

- Many CEOs have clearly articulated that potential catastrophic incidents are keeping them awake at night.
- PSM is a way improve performance of operations and the triple bottom line: people, profit, planet.

Pressure of Performance

- Recent incidents continues to keep PSM as the central focus to improve effective implementation

Skills shortage

- Aging workforce, skill shortage and investigation of catastrophic process incidents due to human errors has increased the need for process safety skill and competency development.

Costs of Incidents

DU PONT

Average Annual Property Damage ($MM) from Top 100 Incidents in Global Oil/Gas Industry

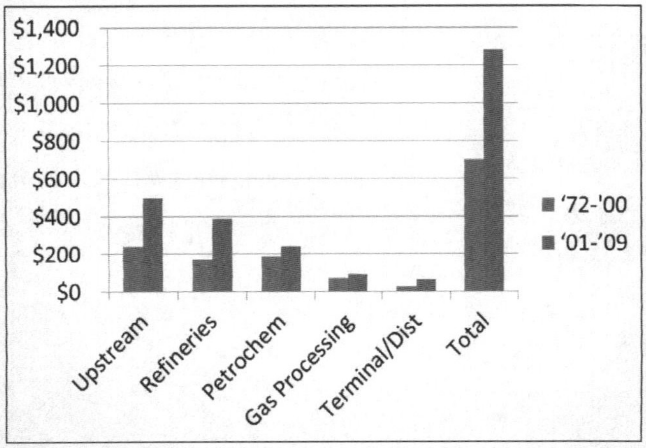

Source: Marsh report

Critical Success Factors

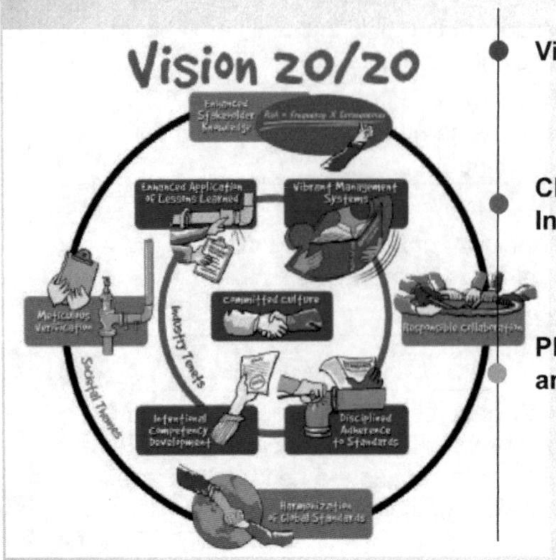

- Visible Leadership Commitment
- Clear Business Case, Strategy and Integration
- Platform to drive People Engagement and Culture transformation

Process Safety is driven by top Leadership Commitment

Ellen Kullman,
Chairman and CEO of DuPont

On Committed Culture

"If your plant sites and your businesses see [PSM] as an important aspect of what [you] value, then they will take that and make it their values as well and will operate with the highest standards of PSM."

Leaders Standard Work

"Committed to ZERO"

Leader Standard Work Component	Operational Discipline LSW	Expected Actions	Supporting Tools	Frequency
Performance dialogues & prioritization	Linewalks	Engagement with shop floor, identification of gaps and removal of barriers, recognition, support of initiatives	Performance dialogue agenda, Visual Board, KPI's, Top 3	Weekly (at least)
Problem solving				
Coaching & delegation				
Process management				
Capability Building Development				

Integrated Management Processes and Systems to drive Sustainable Performance

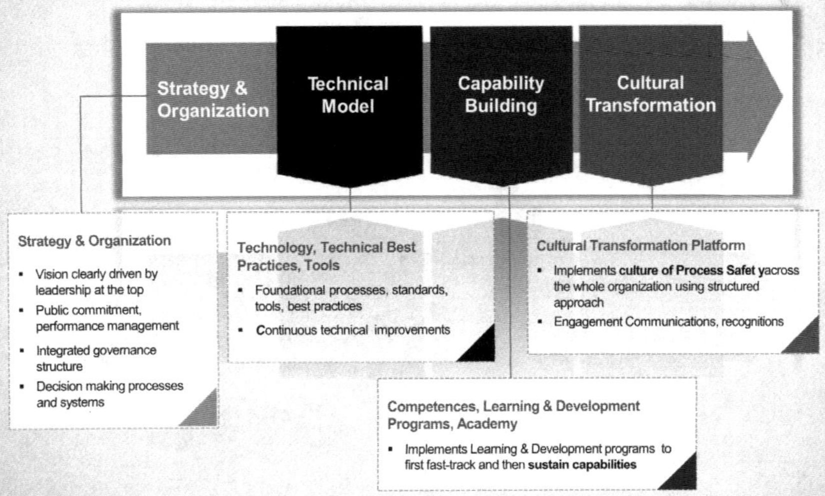

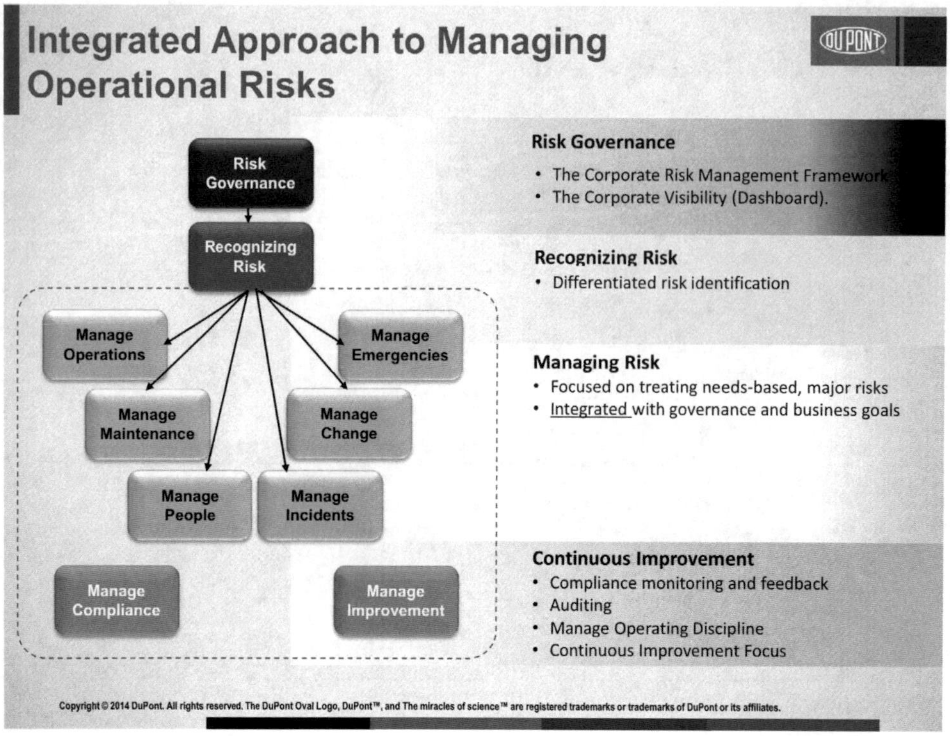
Integrated Approach to Managing Operational Risks
DUPONT
Risk Governance
Recognizing Risk
Manage Operations
Manage Emergencies
Manage Maintenance
Manage Change
Manage People
Manage Incidents
Manage Compliance
Manage Improvement
Risk Governance
• The Corporate Risk Management Framework
• The Corporate Visibility (Dashboard).
Recognizing Risk
• Differentiated risk identification
Managing Risk
• Focused on treating needs-based, major risks
• Integrated with governance and business goals
Continuous Improvement
• Compliance monitoring and feedback
• Auditing
• Manage Operating Discipline
• Continuous Improvement Focus
Copyright © 2014 DuPont. All rights reserved. The DuPont Oval Logo, DuPont™, and The miracles of science™ are registered trademarks or trademarks of DuPont or its affiliates.

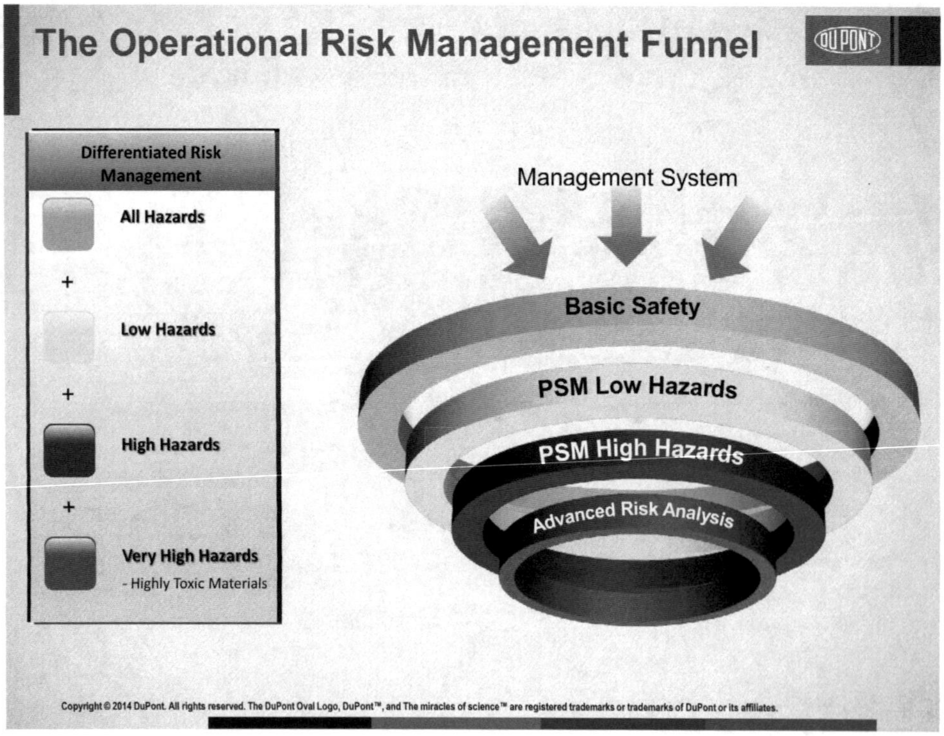
The Operational Risk Management Funnel
DUPONT
Differentiated Risk Management
All Hazards
+
Low Hazards
+
High Hazards
+
Very High Hazards
- Highly Toxic Materials
Management System
Basic Safety
PSM Low Hazards
PSM High Hazards
Advanced Risk Analysis
Copyright © 2014 DuPont. All rights reserved. The DuPont Oval Logo, DuPont™, and The miracles of science™ are registered trademarks or trademarks of DuPont or its affiliates.

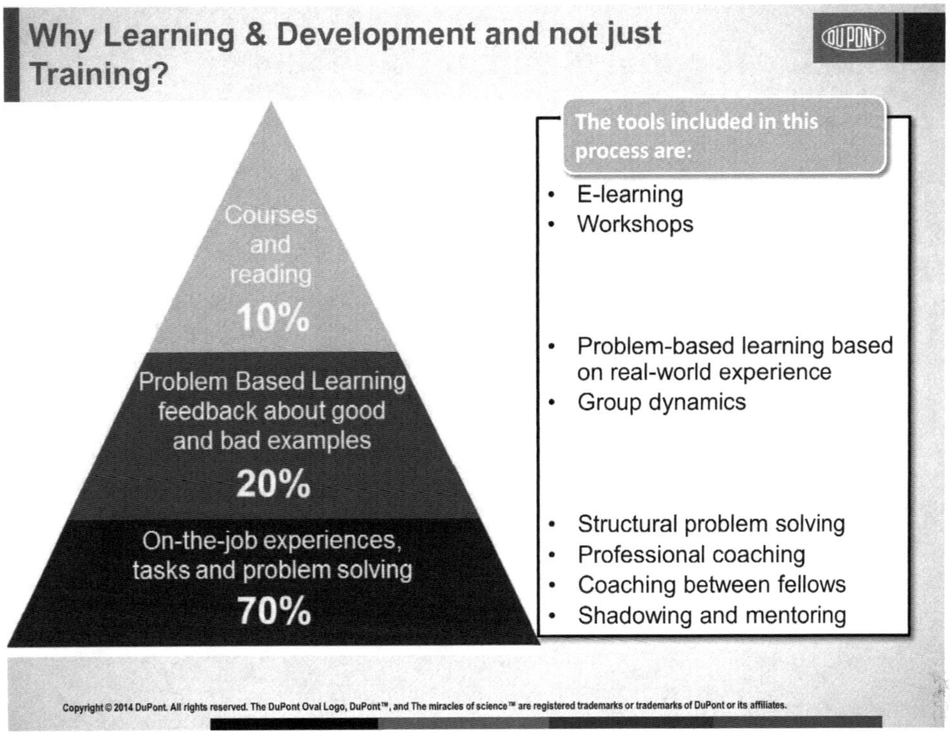
Why Learning & Development and not just Training?
Courses and reading
10%
Problem Based Learning feedback about good and bad examples
20%
On-the-job experiences, tasks and problem solving
70%
The tools included in this process are:
• E-learning
• Workshops
• Problem-based learning based on real-world experience
• Group dynamics
• Structural problem solving
• Professional coaching
• Coaching between fellows
• Shadowing and mentoring
Copyright © 2014 DuPont. All rights reserved. The DuPont Oval Logo, DuPont™, and The miracles of science™ are registered trademarks or trademarks of DuPont or its affiliates.

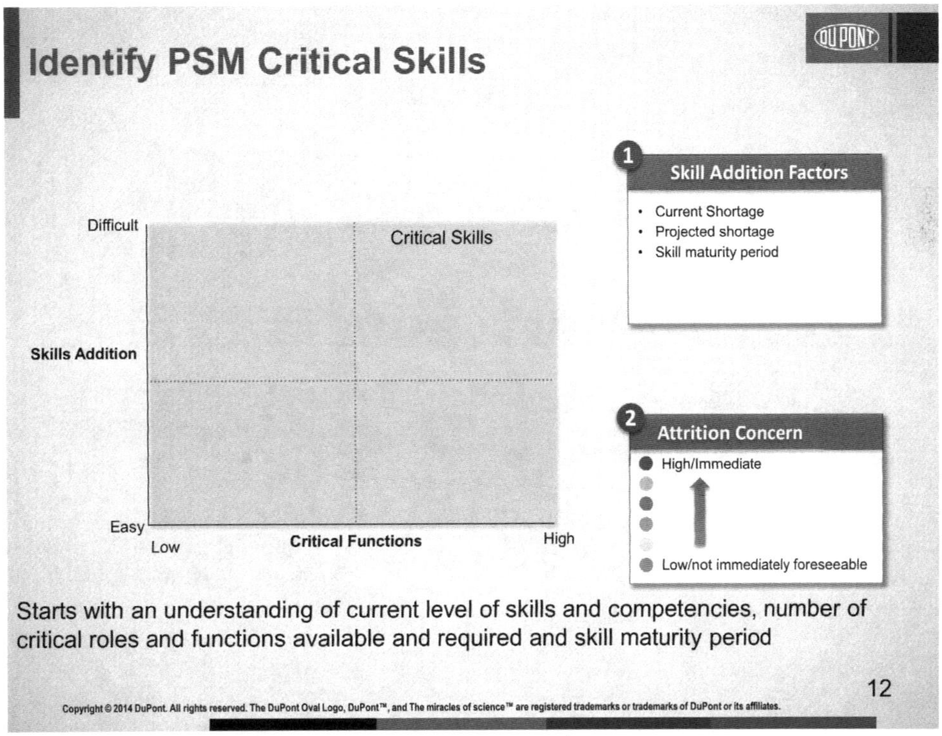
Identify PSM Critical Skills
Difficult
Critical Skills
Skills Addition
Easy
Low
Critical Functions
High
1 Skill Addition Factors
• Current Shortage
• Projected shortage
• Skill maturity period
2 Attrition Concern
High/Immediate
Low/not immediately foreseeable
Starts with an understanding of current level of skills and competencies, number of critical roles and functions available and required and skill maturity period
12
Copyright © 2014 DuPont. All rights reserved. The DuPont Oval Logo, DuPont™, and The miracles of science™ are registered trademarks or trademarks of DuPont or its affiliates.

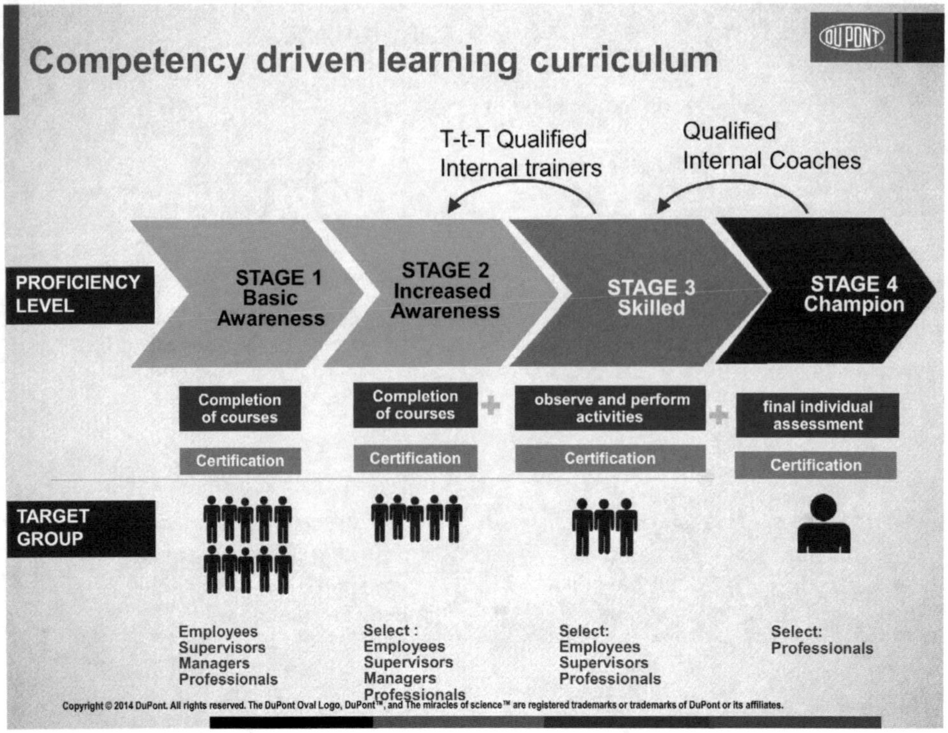
Competency driven learning curriculum
T-t-T Qualified
Internal trainers
Qualified
Internal Coaches
PROFICIENCY LEVEL
STAGE 1
Basic Awareness
STAGE 2
Increased Awareness
STAGE 3
Skilled
STAGE 4
Champion
Completion of courses
Certification
Completion of courses
+
observe and perform activities
Certification
+
final individual assessment
Certification
TARGET GROUP
Employees
Supervisors
Managers
Professionals
Select :
Employees
Supervisors
Managers
Professionals
Select:
Employees
Supervisors
Professionals
Select:
Professionals
Copyright © 2014 DuPont. All rights reserved. The DuPont Oval Logo, DuPont™, and The miracles of science™ are registered trademarks or trademarks of DuPont or its affiliates.

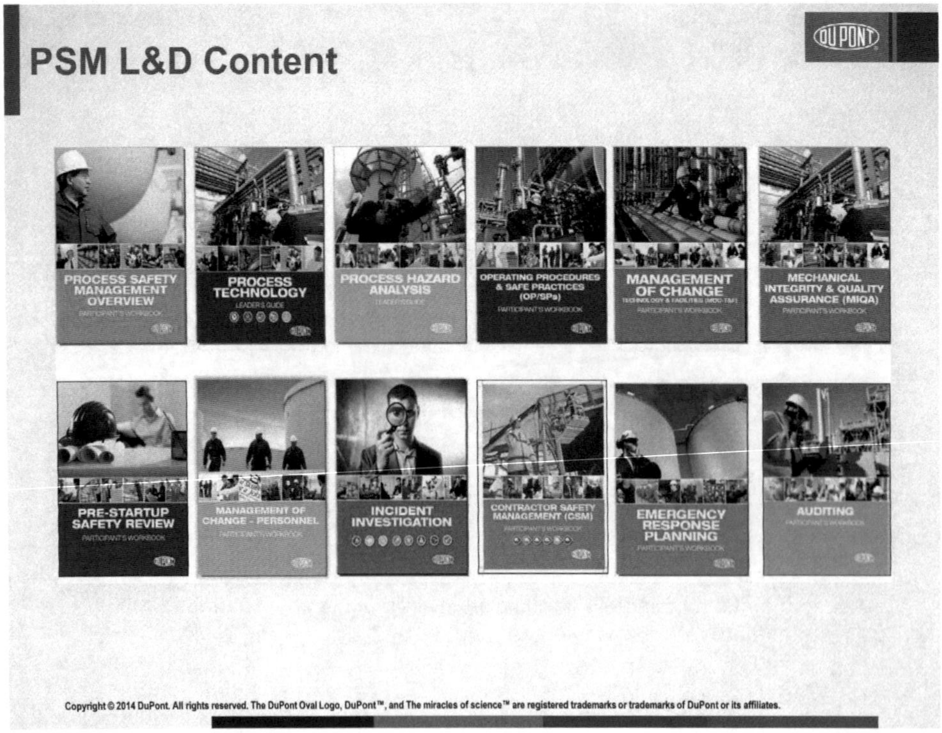
PSM L&D Content
PROCESS SAFETY MANAGEMENT OVERVIEW
PARTICIPANT'S WORKBOOK
PROCESS TECHNOLOGY
LEADER'S GUIDE
PROCESS HAZARD ANALYSIS
LEADER'S GUIDE
OPERATING PROCEDURES & SAFE PRACTICES (OP/SPs)
PARTICIPANT'S WORKBOOK
MANAGEMENT OF CHANGE
PARTICIPANT'S WORKBOOK
MECHANICAL INTEGRITY & QUALITY ASSURANCE (MIQA)
PARTICIPANT'S WORKBOOK
PRE-STARTUP SAFETY REVIEW
PARTICIPANT'S WORKBOOK
MANAGEMENT OF CHANGE - PERSONNEL
PARTICIPANT'S WORKBOOK
INCIDENT INVESTIGATION
CONTRACTOR SAFETY MANAGEMENT (CSM)
PARTICIPANT'S WORKBOOK
EMERGENCY RESPONSE PLANNING
PARTICIPANT'S WORKBOOK
AUDITING
Copyright © 2014 DuPont. All rights reserved. The DuPont Oval Logo, DuPont™, and The miracles of science™ are registered trademarks or trademarks of DuPont or its affiliates.

Learning Organization

1. Clarification of Site and Off-Site Roles and Responsibilities
2. Site Annual PSM Reviews
3. Additional Site/Unit Metrics
4. Communication Practices for Area Shift Turnover
5. Alarm Management
6. Fatigue Management
7. Upgrade of Operational Discipline Approach
8. PSI and the use of RAGAGEP's
9. MOC Training and Testing Requirements

Drivers for Improved Alarm Management

- Incident at one site that resulted in the release of toxic materials:
  - –Lacking a procedure to review active alarms and clear disabled alarms before startup of a new campaign.
  - –Pre-startup safety review procedure does not specifically address review of process safety alarm at startup.
- Incident at one site that resulted in the release of hazardous material:
  - –People became habituated to the repeating alarms
  - –No defined process for dealing with alarm condition

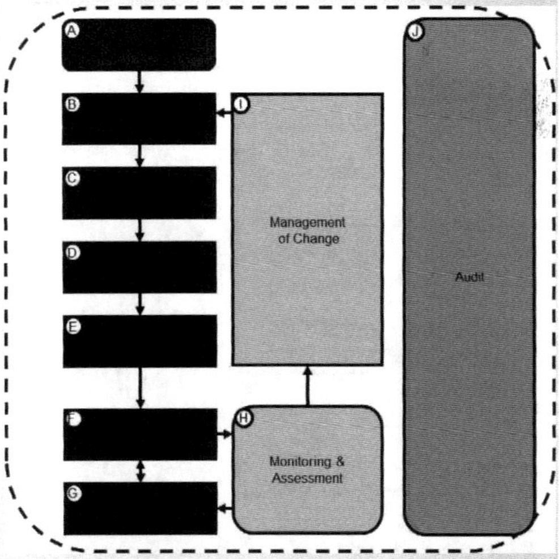

DuPont learnt from these incidents and contributed to development of IEC

Copyright © 2014 DuPont. All rights reserved. The DuPont Oval Logo, DuPont™, and The miracles of science™ are registered trademarks or trademarks of DuPont or its affiliates.

TAKEAWAYS
Visible Leadership Commitment
Clear Business Case, Strategy and Integration
Platform to drive People Engagement and Culture transformation
Copyright © 2014 DuPont. All rights reserved. The DuPont Oval Logo, DuPont™, and The miracles of science™ are registered trademarks or trademarks of DuPont or its affiliates.

DUPONT®
THANK YOU
Srinivasan Ramabhadran
Global Director – Process Safety & Operatina Risk
DuPont Sustainable Solutions
Tel: +65-6586-3644
Email: Ramabhadran.Srinivasan@dupont.com
Copyright © 2014 DuPont. All rights reserved. The DuPont Oval Logo, DuPont™, and The miracles of science™ are registered trademarks or trademarks of DuPont or its affiliates.

Risk Assessment Challenges to 20:20 Vision

Angela E. Summers, PhD, PE
President
SIS-TECH

Dr. Angela Summers
President
SIS-TECH
713-909-2114 (office)
asummers@sis-tech.com

- *Specializes in control and process safety applications*
- *Active participant in many CCPS books*
- *Member of many industrial committees and forums*
- *ISA Fellow*
- *AICHE Fellow*
- *Ph.D. and Engineering Fellow, The University of Alabama*
- *Licensed Professional Engineer in the State of Texas*

2

Risk Assessment Challenges to 20:20 Vision

Angela E. Summers, PhD, PE
President
SIS-TECH

Overview

- Omniscience myth
- Challenges to 20:20 Vision
- On a clear day you can see forever

4

Overview

- Omniscience myth
- Challenges to 20:20 Vision
- On a clear day you can see forever

5

Omniscience Myth

- All events are foreseeable given sufficient analysis and data

6

Omniscience through analysis?

- Every "new" method is a better mousetrap

Omniscience appears unlikely

- More than 1 in 5 events
  - "Organization failing to fully consider potential hazards or causes of component failure"
- In every case, process or equipment design was changed subsequent to the incident

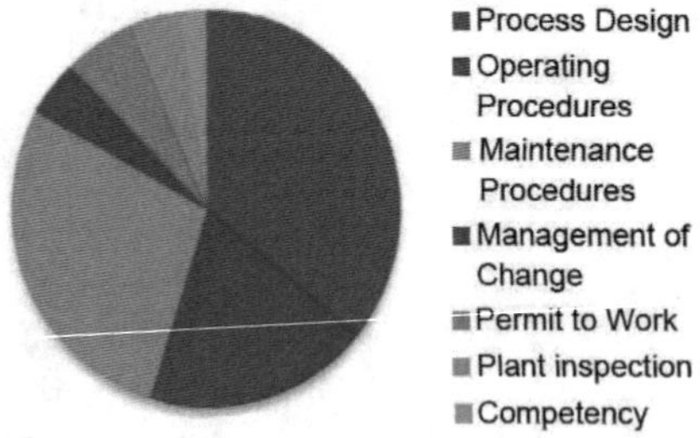

HSE incident analysis

Overview

- Omniscience myth
- Challenges to 20:20 Vision
- On a clear day you can see forever

9

Running before you walk

- "For every complex problem there is at least one solution that is simple, plausible…and wrong."
  - American writer H. L. Mencken
- Implement:
  - Inherent safer design to eliminate event risk
  - Safeguards to reduce event likelihood
  - Response systems to reduce event impact and escalation

10

Onions make me cry

- Barriers are independent as layers of an onion
  - All connected at the root – the management system
- Management system failures affect all layers
  - Culture, engineering, and mechanical integrity practices impact multiple layers

11

'Hole'y moley

- Reason's model addresses deficiencies in barrier performance due to random and systematic faults.
  - Holes open and close dynamically as management systems identify and correct faults
  - The better managed the barriers, the fewer, smaller, and more transient the holes will be

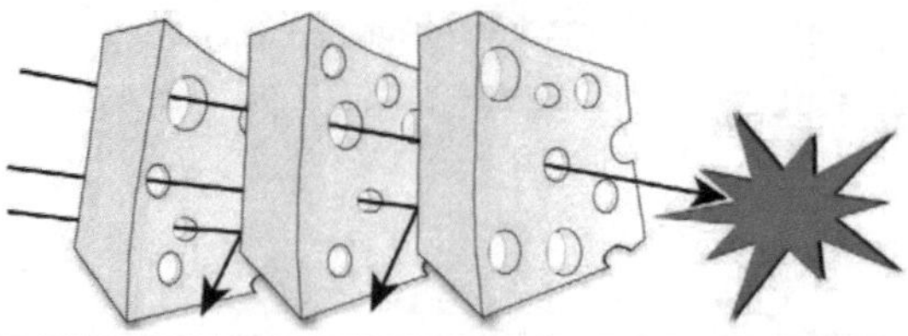

12

Excessive reliance on risk criteria

- Risk is judged based on company risk criteria
- Beware
  - "He uses statistics as a drunken man uses lamp-posts... for support rather than illumination." Andrew Lang

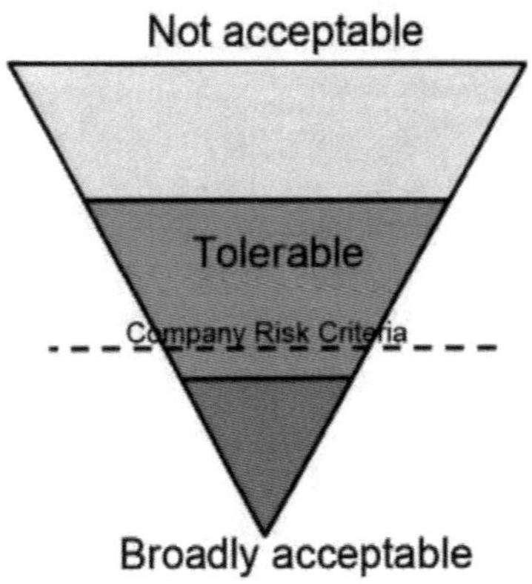

13

No such thing as a perfect 10

- Potential for loss events is directly related to the operational discipline
  - Ensuring the demand rate is less than expected
  - Assumed safeguard reliability is achieved during actual operation

14

Avoid scenario tunnel vision

- Event always seems obvious when the scenario is evaluated
- HSE study
  - 37% of reported incidents resulted from incorrect operator action
  - Root causes - inadequate operating procedures, deficient process design, inadequate supervision, and ineffective management of change.

15

Beware of vogue methods

- Easy to fall into intellectual trap of believing in an analytical perfection in which one “knows” what the risk is.
  - Simple methods are flexible and yield quick results, but repeatability is a challenge
  - Complex methods often present a detailed framework, but are time consuming and error prone

16

Beware hypnotic lure of calculations

- Risk calculations are good for estimating things that can be easily measured.
- Give illusion of accuracy in an environment of high data uncertainty.
  - Causes and scenarios are rarely as mutually exclusive as presumed in the math
  - Most methods do not account for human factors

17

Don't depend on luck factors

- Worse credible scenario is often one where people are present and there is an ignition source for the flammable release.
  - Denial is the greatest enabler – Dr. Clifford Nass, a Stanford professor who pioneered research into how humans interact with technology
  - It is hubris to imagine that we can infallibly prevent a thermodynamically favoured event – P G Urben
  - The fire triangle…should say:

    Air + Fuel = Bang – Trevor Kletz

18

Overview

- Omniscience myth

- Challenges to 20:20 Vision
- On a clear day you can see forever

19

Reasonable Care

- Factors considered in determining whether a company has acted reasonably:
  - Company's care and skill in producing its product
  - Its awareness of the harmful event prior to the incident
  - The activity being performed
  - The specific circumstances that led to the incident
  - ***<u>Whether the company did what it could to prevent the incident's occurrence</u>***

20

Planning for Safe Operation

- Inherent safety
  - Robust vessel and piping design, so process disturbance is tolerable.
- Functional safety
  - Reliable control layer to minimize demand rate.
  - An opportunity for the operator to recover the process
  - A shutdown to prevent loss of containment or other equipment damage
  - Other safeguards to reduce abnormal operation impact
- Emergency response
  - Response systems to reduce event impact and escalation

21

Proof of safe operation

- Benchmark values provide initial basis and rationale for design, but operating history yields:
  - Demand rate
  - Safeguard failure on demand
  - Safeguard failure on test
  - Process safety incident
- Data feedback to the risk analysis process is critical to credible decision-making

22

Safety

- Achieved when there are no loss events
  - Monitor and report actual performance over the life of a process
- KPI
  - **0** loss events
  - **0** demands
  - **0** safety layer failures
  - **0** safety layer out-of-service time

Count von Count, Sesame Street

23

Questions

24

WE FUEL AMAZING

Process Safety Culture in Oil sector

Disclaimer

This material contains forward-looking statements regarding Cairn India and its affiliates, our corporate plans, future financial condition, future results of operations, future business plans and strategies. All such forward- looking statements are based on our management's assumptions and beliefs in the light of information available to them at this time. These forward-looking statements are by their nature subject to significant risks and uncertainties; and actual results, performance and achievements may be materially different from those expressed in such statements. Factors that may cause actual results, performance or achievements to differ from expectations include, but are not limited to, regulatory changes, future levels of industry product supply, demand and pricing, weather and weather related impacts, wars and acts of terrorism, development and use of technology, acts of competitors and other changes to business conditions. Cairn India undertakes no obligation to revise any such forward-looking statements to reflect any changes in Cairn India's expectations with regard thereto or any change in circumstances or events after the date hereof. Unless otherwise stated the reserves and resource numbers within this document represent the views of Cairn India and do not represent the views of any other party, including the Government of India, the Directorate General of Hydrocarbons or any of Cairn India's joint venture partner.

Process Safety & Culture

- Process Safety – control of major accidents

  *A disciplined framework for managing the **integrity of hazardous operating systems and processes** by applying good design principles, engineering, and operating and maintenance practices*

- Culture – the way we do things around here.

  "...the product of the individual and group values, attitudes, competencies and patterns of behaviour that determine the commitment to and the style and proficiency of, an organizations heath and safety programs"

Major accidents

Piper Alpha

Macondo (BP): Deepwater Horizon

Common theme – organisational Safety Culture

Safety culture

- Impacted by country and local cultures
- Safety Culture transcends beyond individuals

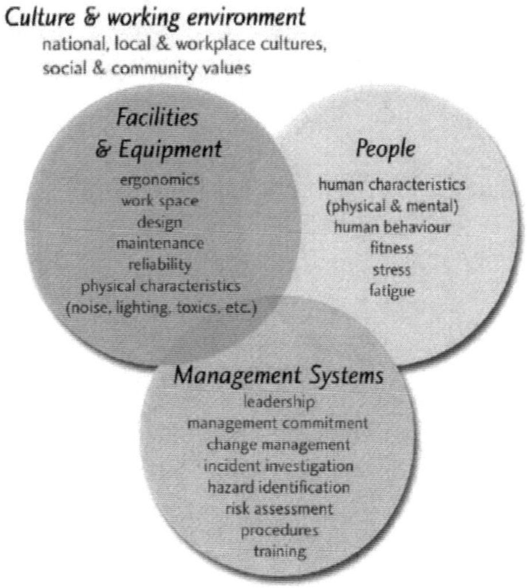

How to measure?

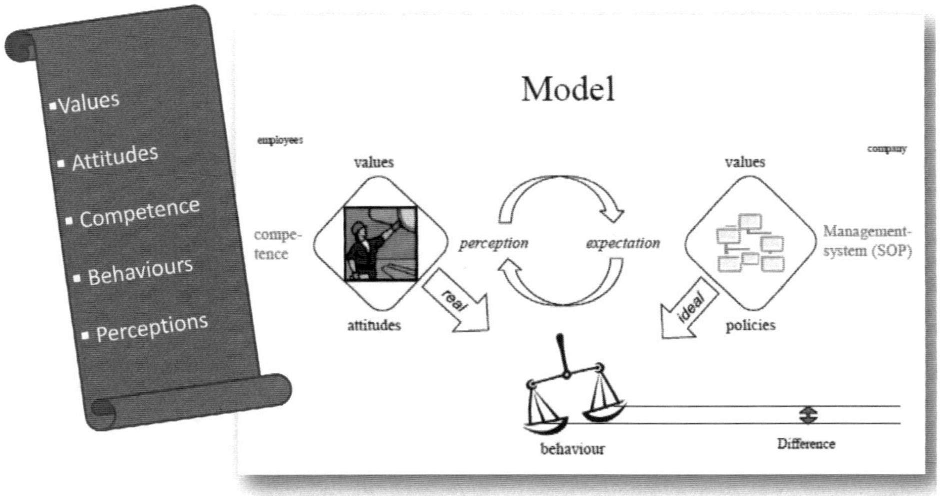

Culture maturity model

Pathological - Lawyers said it was OK! Of course we have incidents it's a dangerous business!

Reactive - We are serious but why don't they do what they are told !

Calculative - We cracked it, lots and lots of audits

Proactive - Management is open but still obsessed with statistics

Generative – Safety is profit centre, Chronic unease

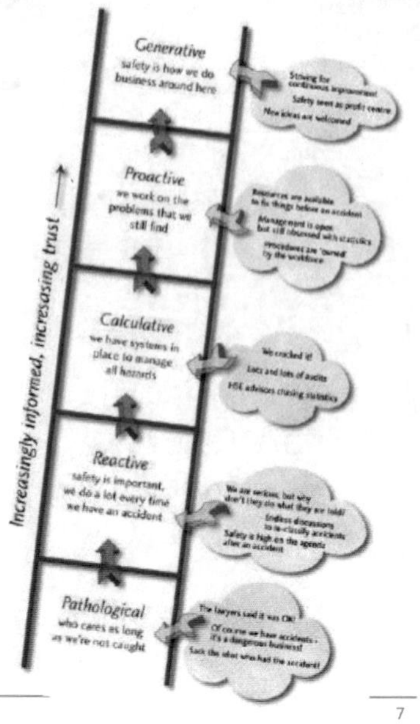

1/13/2015 7

Process safety journey

- Leadership focus
  - MAH awareness
  - KPI development
- Baseline assessment
  - System evaluation
  - Culture assessment
- Course correction after Macondo
  - Inclusion of Drilling
- Cumulative Risk barrier Model

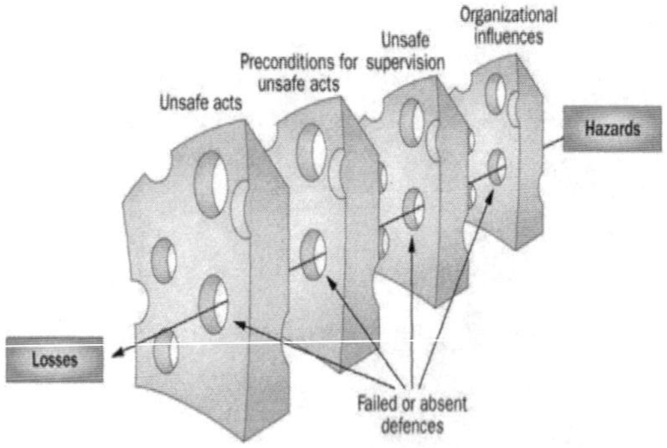

1/13/2015 8

Cairn Journey so far

- Awareness
  - Manager tours
- Systems
  - MOC process
  - PTW
  - Well Integrity
- Baseline assessment

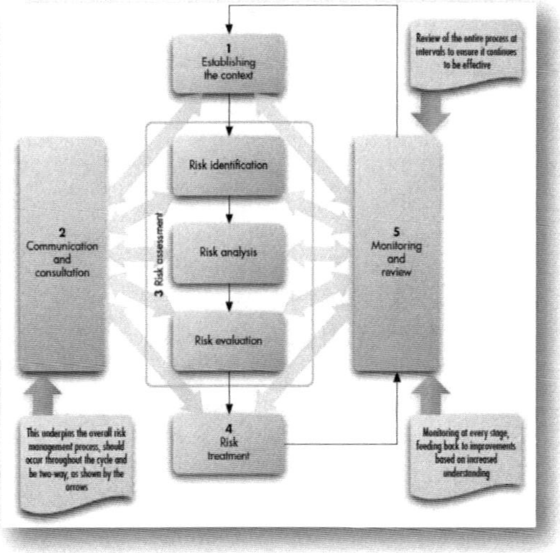

Lagging & leading indicators: 4 Tier Approach

- Based on IOGP / API 754
- **Lagging Indicators:** Tier 1 & 2
  - Accidents already happened
  - Thorough Incident investigation
  - Lesson learnt Programs
- **Leading Indicators:** Tier 3 & 4
  - Pro-active in nature
  - Show evidences of maintenance and improvement of **risk barrier**

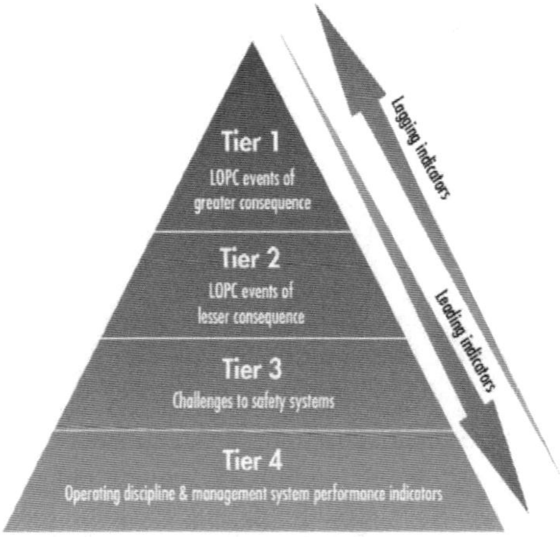

Process Safety Pyramid

Only culture can affect all the 'cheese slices' ... - James Reason

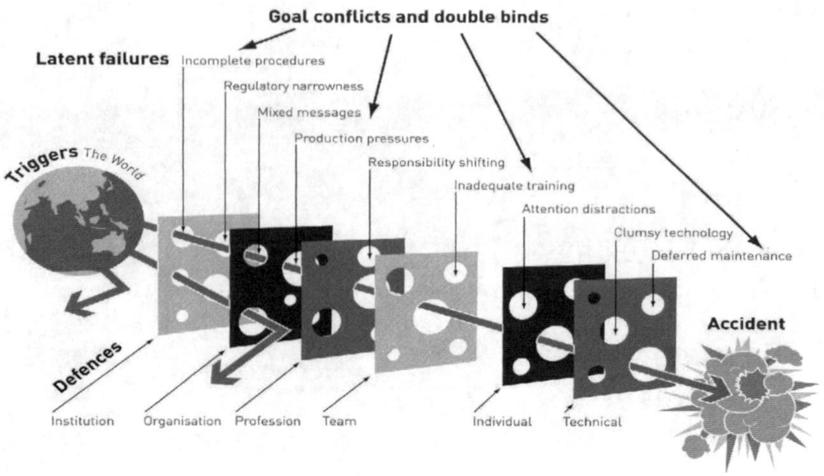

1/13/2015 11

1/13/2015 12

Orchid Chemicals & Pharmaceuticals Ltd.
Shaping A Dream

CHALLENGES IN INFLUENCING PROCESS SAFETY CULTURE IN PHARMACEUTICAL INDUSTRIES

S.Mani & B.Sangeetha
Corp SH&E
Orchid Chemicals & Pharmaceuticals Ltd, Chennai

8th CCPS Asia Pacific Regional Meeting, Mumbai
September 22, 2014

Safety Pause

Video of Aura Reaction Mix transfer Pump Seal Fire (Removed in this attachment)

Mechanical Integrity

Introduction

- Indian Pharmaceutical Industry a low cost producer
- Involves Process Research, Up-scaling to Commercial Production and Formulations
- Products are Oral / Injectable
- Involves very sophisticated equipment and aseptic packaging

Typical API Plant Configuration

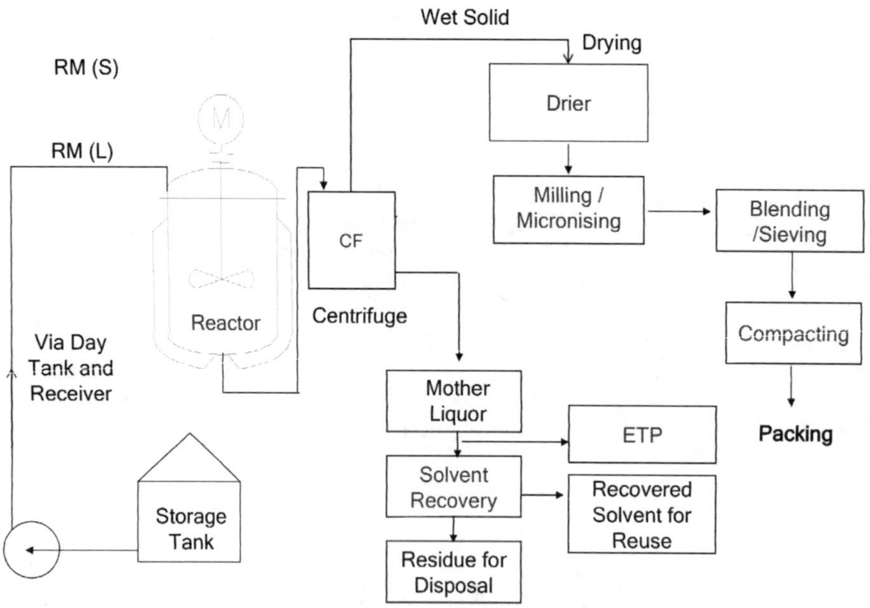

Process Flow Diagram of Formulation Manufacturing Activity

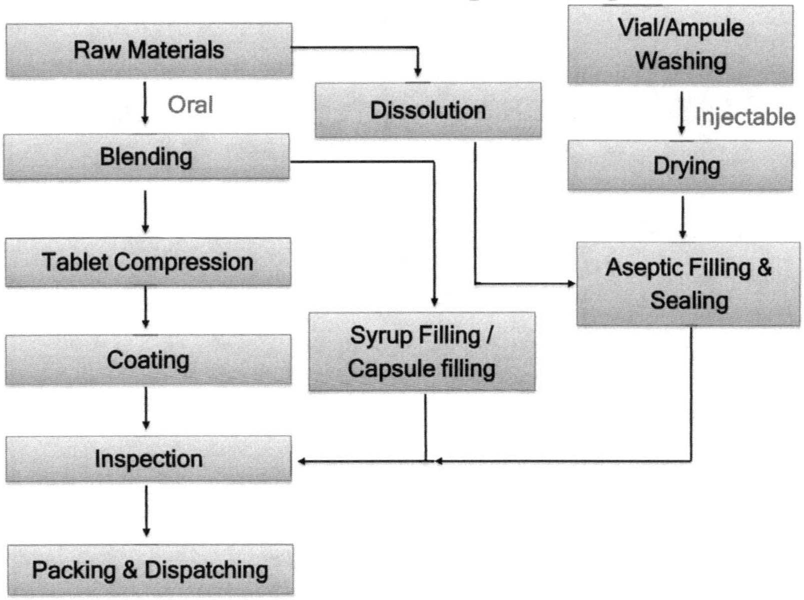

Nature of manufacturing

- Batch processes
- Multi Step, Multi-product
- Low/High temperature technology
- Large quantity of Solvents
- Unit Operations

Culture

"It's what we do when no one is watching!"

Safety Culture – Driving Factors

- Demonstrating visible felt Leadership
- Line Management accountability
- Effective two way Communication
- Willingness to learn & Responsiveness to change
- Manifest attention to work place and process safety
- Intervene "at risk Behaviours" timely
- Strong Observation and Audit Programs
- Foster Mutual trust
- Empower individual to fulfill their responsibility

PSM

Elements of Process Safety Management

Challenges

Challenges - PSM – Technology

- Availability of Process Safety Information
  - ✓ Non availability /Inadequate PSI (PTB+Acetone video to be added)
  - ✓ Confidentiality in "Pharma Process"
  - ✓ Continuous change in Technology
- Gap in Understanding "Why aspects" in Procedures by frontline employees
- Recognition of hazards (reactive) – deformation of plastic tray in drying oven ; Spark in Steam Hose video (video to be added)
- Inadequate Risk Assessment

Challenges - PSM – Facility

- Quality Assurance
  - ✓ Sound Systems in place
- Frequent changes in facility due to product campaign
- Mechanical Integrity
  - ✓ Compromise on Preventive maintenance
- Shortcoming in risk assessment while carrying out changes in facility / equipment spares

Challenges - PSM – Personnel

- Attrition rate
- Identification of training need and methodology
- Training effectiveness
- Recognition and reporting of near miss
- Thorough investigations to find the exact root causes of incident
- Sharing of Lessons learned among pharma industries from Incidents
- Equipping people during promotions / transfers
- Observation by leaders and line mangers to demonstrate the care for individual

Influencing People

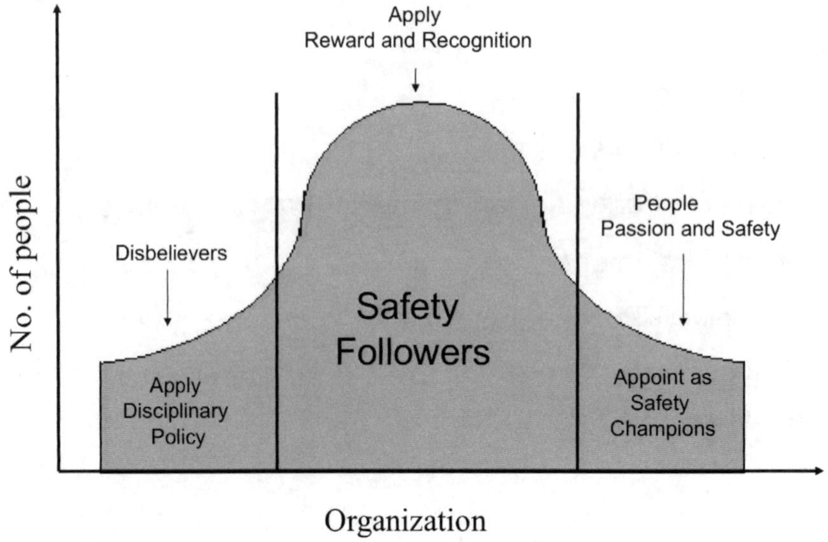

Source:DuPont

Building a Safety Culture

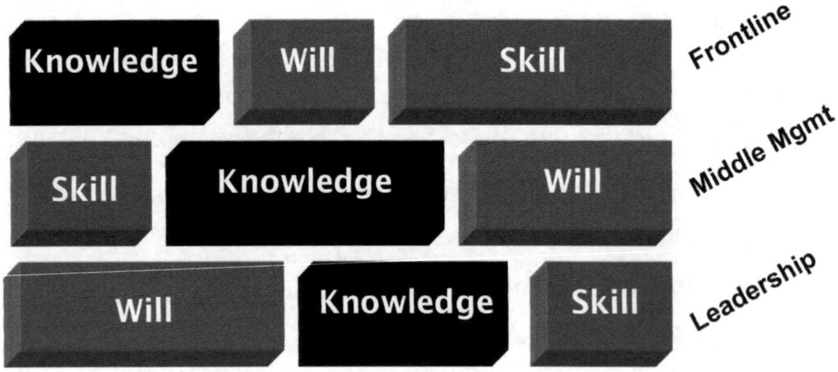

Felt Leadership - starts at the top

Felt leadership is one of the top differentiators between a good and great safety culture.

10 Leadership Principles:

1. **Be visible to the organization**
2. **Be relentless about time with people**
3. Recognize your role as teacher/trainer
4. **Develop your own safety functioning skills and pass them along to the organization**
5. Behave and lead as you desire others to do
6. **Maintain a self-safety focus**
7. Confirm and reconfirm safety as the #1 value
8. **Place continuous emphasis and clarity around safety expectations**
9. Show a passion for ZERO injuries, illnesses, and incidents
10. **Celebrate and recognize ZERO successes**

Felt Leadership

Safety Excellence is a

CULTURAL TRANSFORMATION

that requires

Felt Leadership

Safety Excellence --> Cultural Excellence

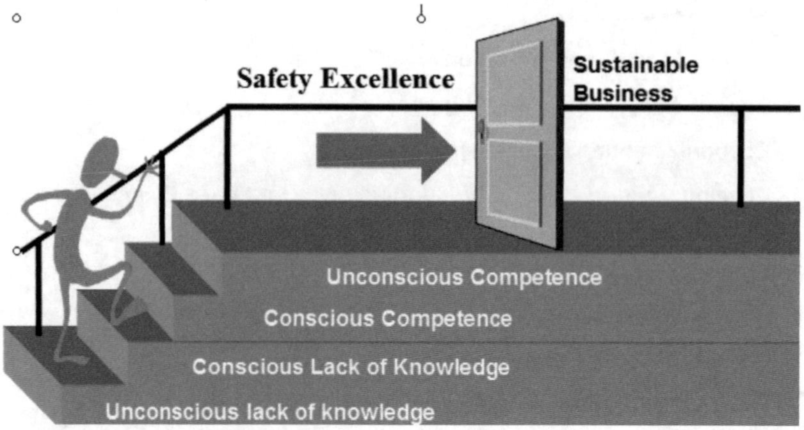

mani@orchidpharma.com

Click to go back

TATA CHEMICALS LIMITED

PROCESS SAFETY
Culture Enablers

S Bhasker Kumar 15th Dec 2014

Tata - Leadership with Trust

- Gross revenue > $103 Billion with 67% generated in geographies other than India
- Operations in more than 100 countries across six Continents,
- Total Number of Employees >580,000
- Conglomerate of over 100 companies in 7 Business Sectors:
  - Materials
  - Energy
  - ***Chemicals***
  - Services
  - Consumer Products
  - IT & Communications
  - Engineering Products & Services

TATA CHEMICALS LIMITED 2

Tata Chemicals – Global Presence

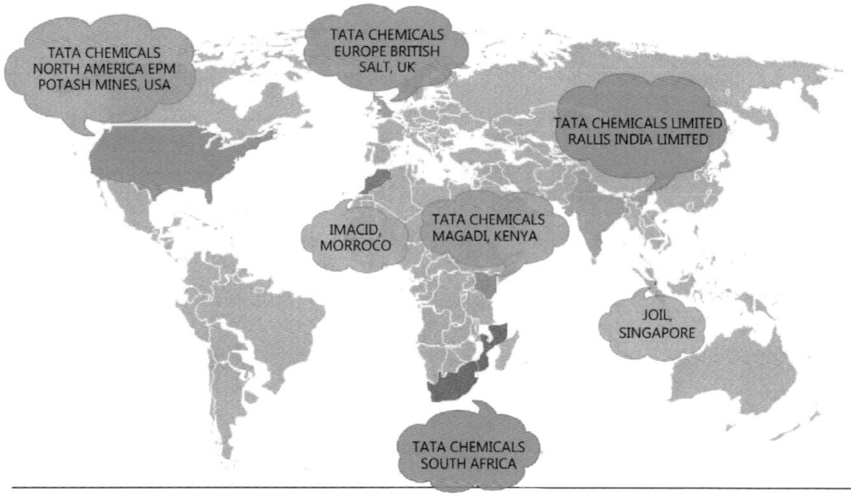

TATA CHEMICALS LIMITED

3

Tata Chemicals- Mission, Vision & Values

Mission

Serving Society through Science

Vision

We shall be amongst premier chemical companies by:

- Leveraging Science to deliver new and innovative offerings
- Enhancing value to our customers
- Delivering Superior Returns to our shareholders
- Leading in corporate sustainability
- Nurturing innovation, learning through diversity and team work amongst employees

Values

Integrity, **Safety**, Excellence, **Care** & Innovation

TATA CHEMICALS LIMITED

4

Tata Chemicals Works in India

TATA

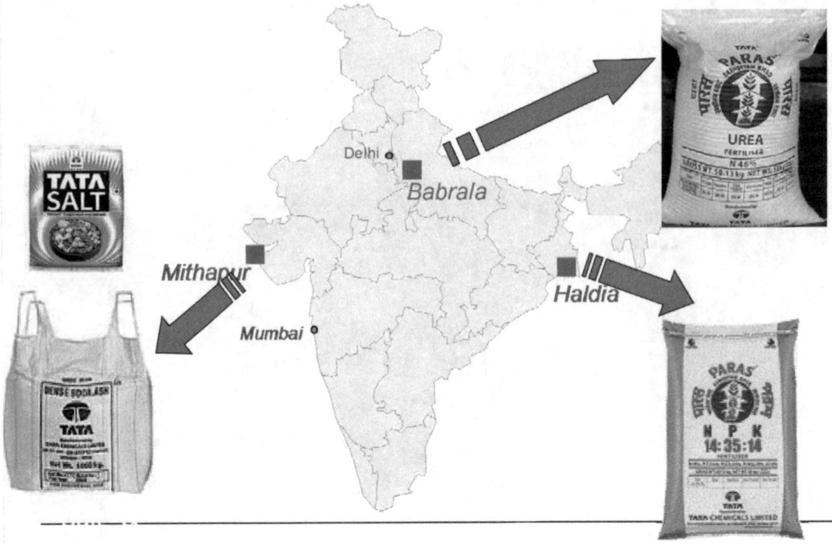

TATA CHEMICALS LIMITED

5

Tata Chemicals Ltd - Babrala Site

TATA

- Site area : 1520 acres
- Plants : Ammonia, Urea, O&U, Product Packaging and Customized Fertilizers
- Plant Capacities
  - Ammonia 2000 MTPD
  - Urea 3500 MTPD
  - Customized Fert. 400 MTPD
- Technology
  - Urea : Snamprogetti, Italy
  - Ammonia : Haldor Topsoe, Denmark
  - CF : A J Saket, USA

TATA CHEMICALS LIMITED

Urea Manufacturing Process

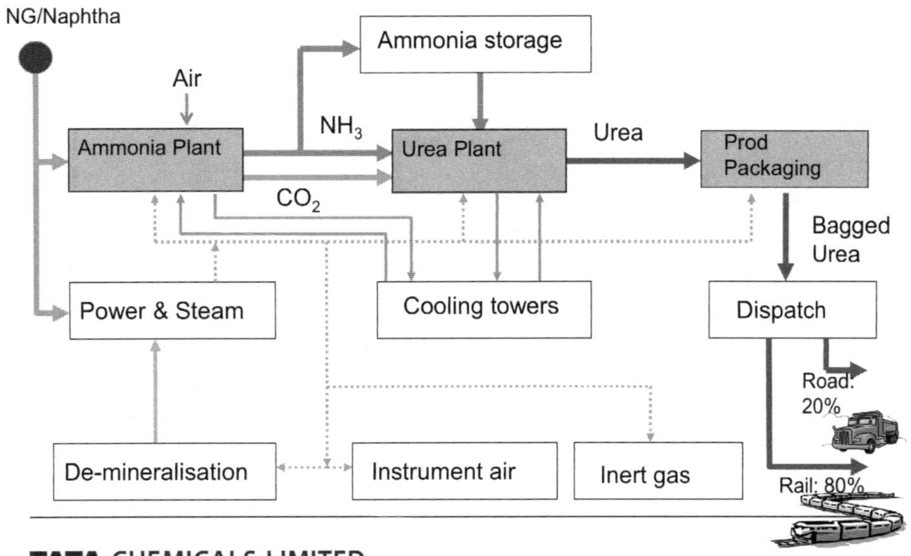

TATA CHEMICALS LIMITED

7

Operational Excellence through Processes & People

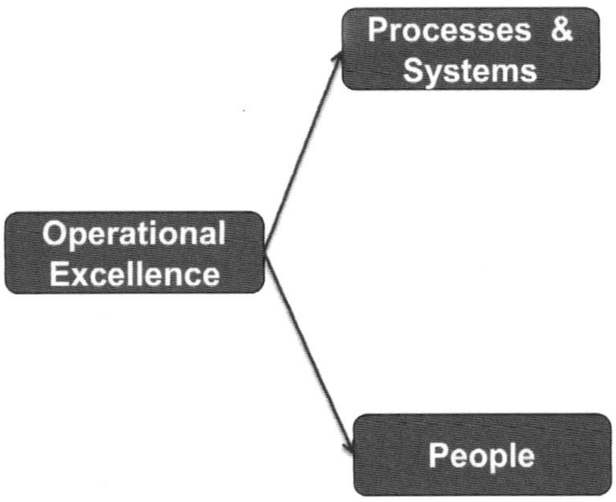

- Strong Process & Systems orientation
- Focus on excellence

- Doing right for the first time
- People the most valuable assets
- Focus on capability building
- Open and equal opportunity climate
- Knowledge Management
- Culture of learning & Sharing

TATA CHEMICALS LIMITED

8

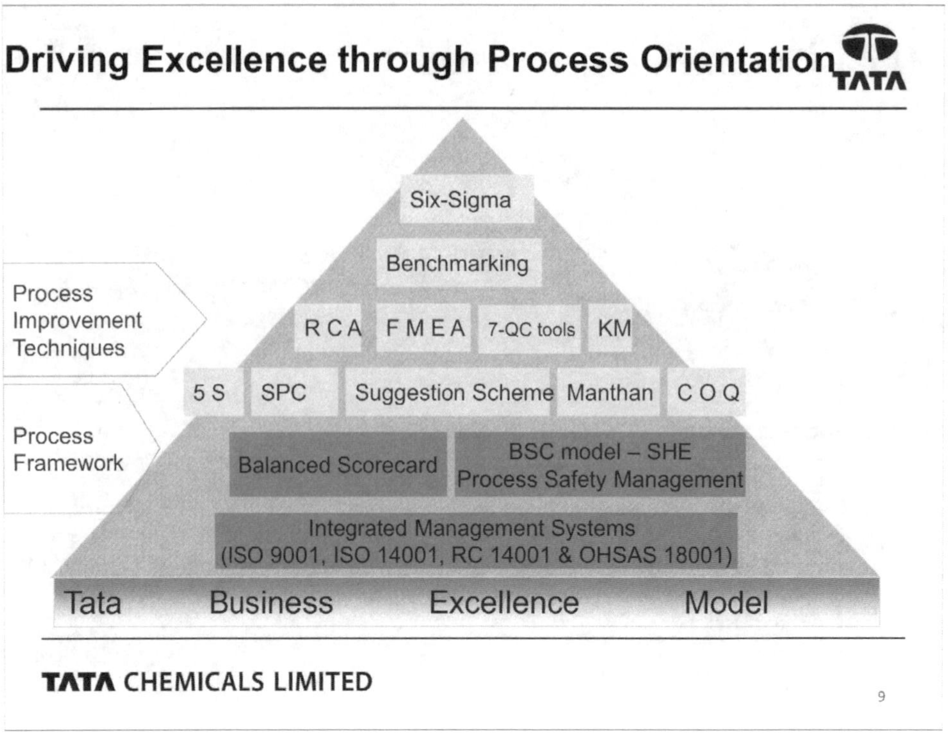
Driving Excellence through Process Orientation
TATA
Six-Sigma
Benchmarking
Process Improvement Techniques
R C A
F M E A
7-QC tools
KM
5 S
SPC
Suggestion Scheme
Manthan
C O Q
Process Framework
Balanced Scorecard
BSC model – SHE Process Safety Management
Integrated Management Systems (ISO 9001, ISO 14001, RC 14001 & OHSAS 18001)
Tata Business Excellence Model
TATA CHEMICALS LIMITED
9

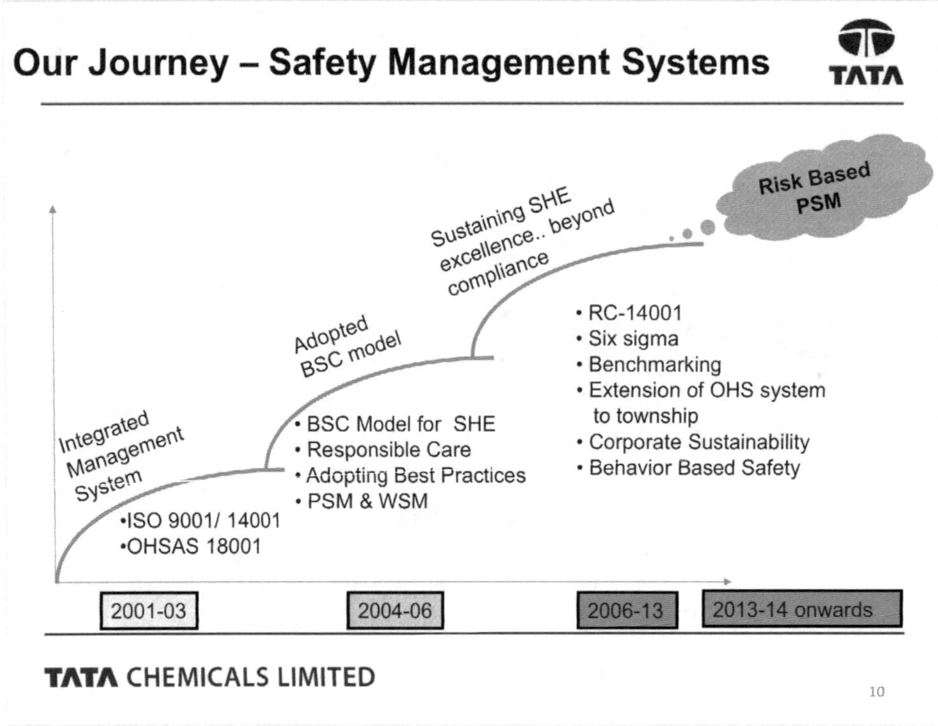
Our Journey – Safety Management Systems
TATA
Risk Based PSM
Sustaining SHE excellence.. beyond compliance
Adopted BSC model
Integrated Management System
•ISO 9001/ 14001
•OHSAS 18001
• BSC Model for SHE
• Responsible Care
• Adopting Best Practices
• PSM & WSM
• RC-14001
• Six sigma
• Benchmarking
• Extension of OHS system to township
• Corporate Sustainability
• Behavior Based Safety
2001-03
2004-06
2006-13
2013-14 onwards
TATA CHEMICALS LIMITED
10

In spite of running process safety programs in many facilities, major process incidents continue to occur??

TATA CHEMICALS LIMITED

11

The Key is

Process Safety Culture

Which is the Pillar

TATA CHEMICALS LIMITED

12

Culture: Two aspects

TATA

- Something an organisation is: shared values and beliefs.
- Something an organisation has: structures, practices, systems.
- Changing practices easier than changing values and beliefs.

TATA CHEMICALS LIMITED

13

Process Safety Culture

TATA

The UK HSE defines safety culture as

" . . . the product of the individual and group values, attitudes, competencies and patterns of behavior that determine the commitment to the style and proficiency of an organization's health and safety programs."

TATA CHEMICALS LIMITED

14

Process Safety Culture

"The way organization behaves when no one is watching"

TATA CHEMICALS LIMITED

15

Process Safety Culture

- 'Safety culture' is a subset of the overall company culture.
- 'safety culture' may be used to refer to the inclination of employees to comply with rules or act safely or unsafely
- However, culture and style of management is even much more deeper
  - e.g. a natural, unconscious bias for production over safety, or a tendency to focussing on the short-term and being highly reactive, etc.

TATA CHEMICALS LIMITED

16

Types of Process Safety Culture

TATA

Organic	Dynamically responding to risk. Quick to learn and improve
Proactive	Learning from incidents
Certificate	Implement system on paper
Reactive	React after an incident
Denial	What has not happened will never happen

TATA CHEMICALS LIMITED

17

Some barriers to cultural progression

TATA

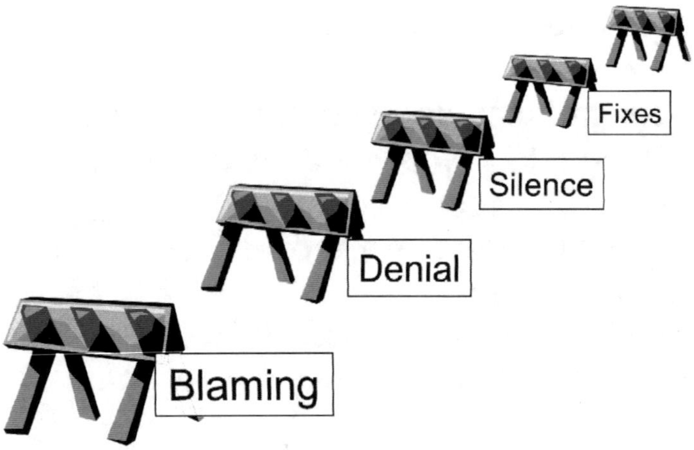

TATA CHEMICALS LIMITED

18

PSM IS A MINDSET CHANGE

TATA

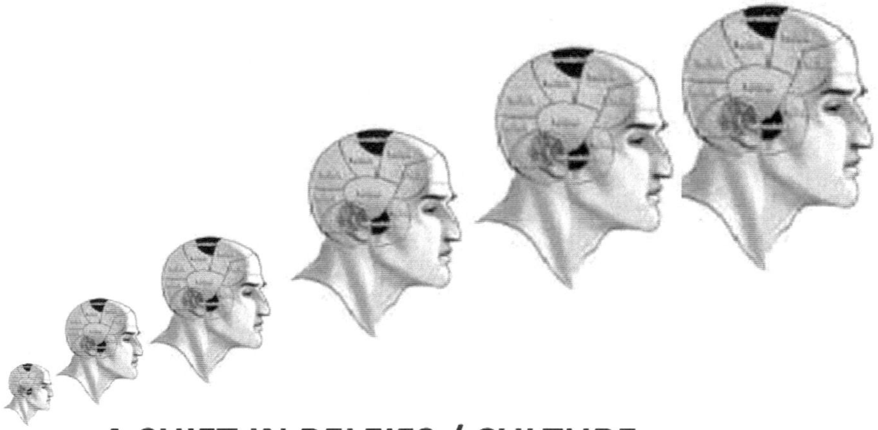

A SHIFT IN BELEIFS / CULTURE

TATA CHEMICALS LIMITED

19

Actively Caring and the Safety Triangle

TATA

Environment

Making sure needed equipment is available. Posting warning signs, housekeeping, cleaning other's work area.

Person

Sharing skills and knowledge with each other. Listening, helping in a crisis, recognizing team member contributions.

Behavior

Observing co-workers, giving feedback, modeling behavior.

Often neglected in traditional safety approaches. Little or no feedback on or encouragement of safe behavior.

TATA CHEMICALS LIMITED

20

Actively Caring Increases Effectiveness

TATA

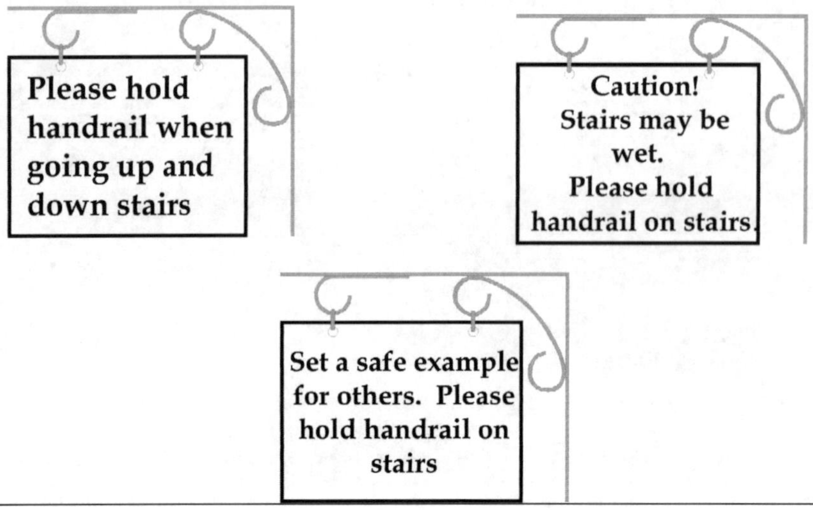

TATA CHEMICALS LIMITED

21

Values, Intentions and Behaviors

TATA

Cautioning co-workers about performing unsafe acts

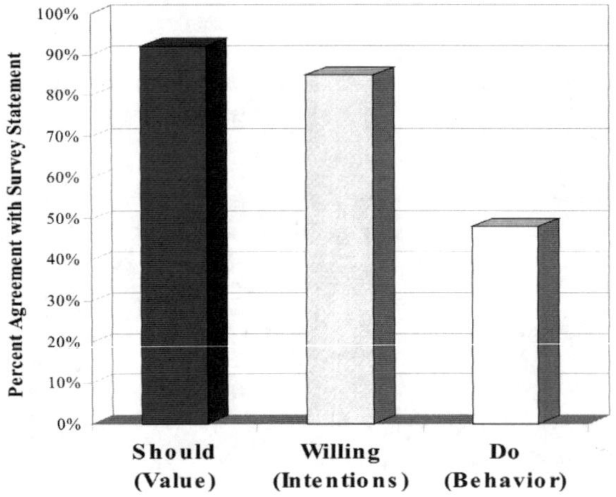

TATA CHEMICALS LIMITED

22

Developing Safe Habits

TATA

I no longer think about it; I know it is right and now it is a habit

Unconsciously Competent

I know I am doing this right because I'm following the procedure

Consciously Competent

Consciously Incompetent

I know there's a better way, I need to learn it."

Unconsciously Incompetent

I didn't know there was a better way

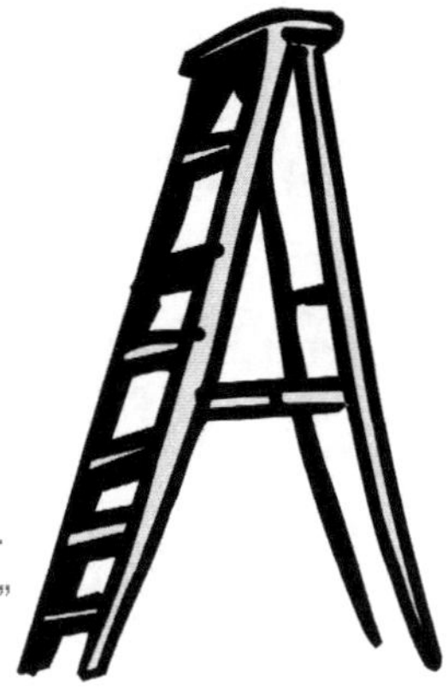

TATA CHEMICALS LIMITED

23

Process Safety Culture Enablers

TATA

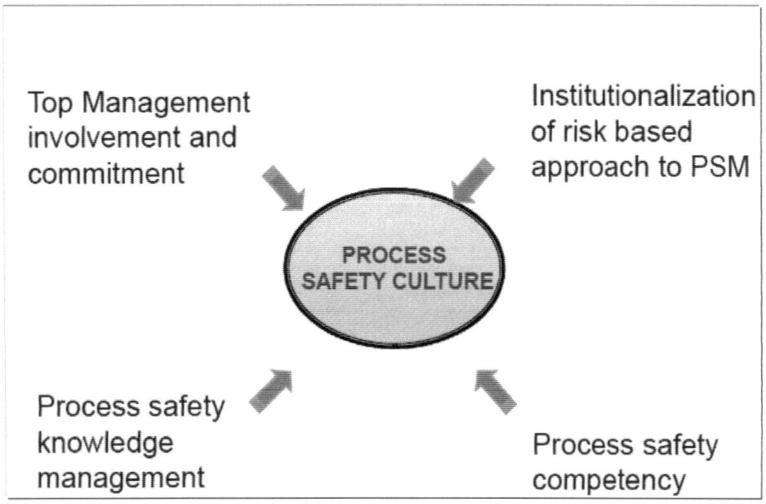

TATA CHEMICALS LIMITED

24

Top Management Involvement & Commitment

TATA CHEMICALS LIMITED

25

Top Management Involvement & Commitment

- Personal involvement of senior management
- Process safety versus production, cost etc.
- Must "Walk the Talk" and "Lead by Example"
- Felt Leadership: demonstration of visible and credible commitment
  - Appear regularly on the 'shop floor', talk about OHS and process safety
  - visibly demonstrate their commitment by their actions
  - perceived as 'sincerely committed to safety'

TATA CHEMICALS LIMITED

26

Top Management Involvement & Commitment

Other examples of required leadership behaviours:

- Maintaining a "Sense of Vulnerability"
- Avoiding "Normalization of Deviation"
- Receptiveness to bad news by top management
- Ensuring process safety competency at decision making levels and operating levels
- Focus on measuring process safety performance rather than only relying TRIR

TATA CHEMICALS LIMITED

27

Root Causes

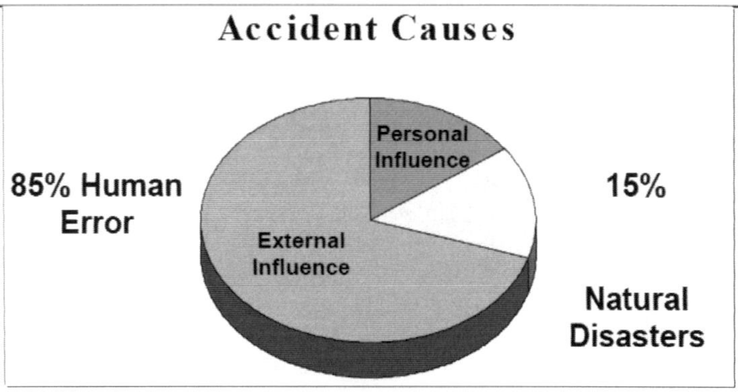

Accidents

↑

Human Errors

↑

Human Factors (weak management systems / culture)

TATA CHEMICALS LIMITED

28

Top Management Involvement & Commitment

- **Ensures Learning from Incident:** Process incidents and near misses are reported and investigated to their root causes.
  - root causes mostly lie within organisational issues or human factors (system & culture)
  - **Human factors** are not the same as **human errors** and are organisational or technical factors which led people to make mistake
  - Top management to avoid getting stuck up in the **"Human errors"** and focus on "**Human factors"**

TATA CHEMICALS LIMITED

29

Steps for visible commitment & involvement by sr. management

- Encourage visual displays and communication on 9 golden safety rules
- **Senior management** to visit plant **at least once a week**, discuss one golden rule and relate with any job going on and share any related incident with plant operators and maintenance technicians.
- Head of departments to participate in morning meeting with technicians/maintenance contract workers **at least twice a week** and talk about safety
- During training class, site head to address the class at the beginning **at least once a month**

TATA CHEMICALS LIMITED

30

Steps for visible commitment & involvement by sr. management - contd.

- Improve effectiveness of consequence management system. (display of cardinal rules, communication about the same, circulation of consequence matrix among HODs and adherence)
- During any major jobs, shutdown or turnaround, Sr. management must talk about safety first, during planning, before starting or during the job (more emphasis to be laid on safety)
- Personal involvement of Sr. Management in reporting near misses, process incidents, BBS observations
- Participate and encourage their dept colleagues also to participate in all SHE related meetings, activities and celebrations

TATA CHEMICALS LIMITED

31

Steps for visible commitment & involvement by sr. management - contd.

- Sr. Mgt (i.e HOD's + Executives) should be **role models** in their conduct with respect to safety procedures, use of PPEs, adherence to timings etc which slowly becomes a culture

- Need to spread / inculcate culture of **"Duty of Care"** which can have multiplying effect in building safety culture.

TATA CHEMICALS LIMITED

32

Risk Based Decision Making for PSM

TATA CHEMICALS LIMITED

33

Institutionalization of Risk Based Decision Making

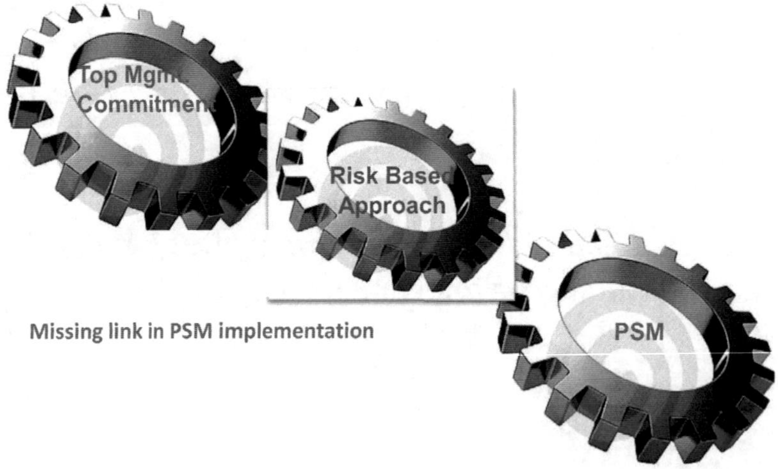

TATA CHEMICALS LIMITED

34

Risk Based Approach

- Focuses on operation and maintenance risk in the process plant
- Risks are minimized As Low As Reasonably Practical **(ALARP)**
- Institutionalizes risk based decision making
  - Uniform understanding of whether risk are acceptable or not (organization wide alignment in the perception and control of process risks)
  - Communication of process risks to the appropriate levels
  - Process safety is considered in a systematic way and does not get overridden even during times of pressure

TATA CHEMICALS LIMITED

35

As Low As Reasonably Practical (ALARP)

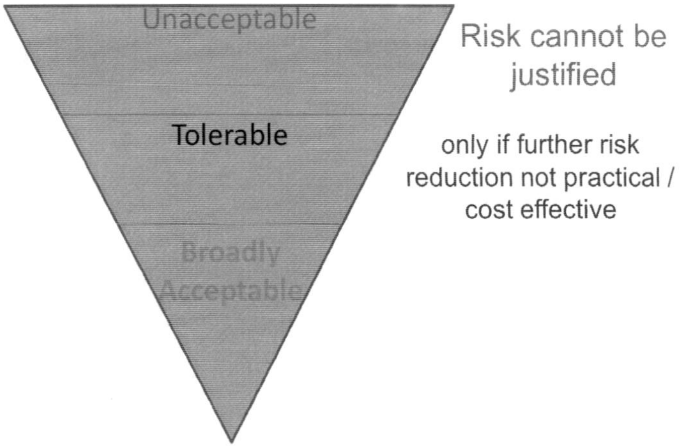

TATA CHEMICALS LIMITED

36

Operations & Maintenance Risk Matrix

TATA

- Operations and maintenance risk matrix aligned with overall risk management strategy of TCL has been developed
- The risk matrix is based on the **As Low As Reasonably Practical (ALARP)** philosophy.
- All decisions that could create **process safety exceptions** shall be taken through the operations and maintenance risk matrix to assess the risk level

TATA CHEMICALS LIMITED

37

OPERATIONAL AND MAINTENANCE RISK MATRIX

PROBABILITY ↑					
Frequent 1 in 2 years or less					
Probable 1 in 5 years				HIGH	
Occasional 1 in 10 years			MEDIUM		
Remote 1 in 20 years		LOW			
Improbable 1 in 50 years					

IMPACT ⟶

People	First aid case (FAC)	Medical treatment cases or Restricted work case (MTC or RWC)	Lost Work Day cases (LWC)	Single Fatality or Permanent disability	Multiple fatalities
Assets	Slight damage	Minor damage	Local damage	Major damage	Extensive damage
Environment	Slight effect	Minor effect	Localised effect	Major effect	Massive effect
Reputation	Slight impact	Minor impact	Regional impact	National impact	World wide impact

38

Process Safety Exception

Any management decision that does not align with the major objective of Process Safety Management (which is to prevent unwanted releases of hazardous chemicals especially into locations that could expose employees and others to serious hazards)

Process Safety Exception

Few examples:

- Not releasing equipment for planned inspection schedule
- Extending service of an equipment/pipeline etc already recommended by inspection to be repaired/replaced
- Postponing any Pre Startup Safety review (PSSR) recommendation but going ahead with commissioning of the equipment
- Extending process incident investigation recommendations implementation dates.
- Extending Process Hazard Analysis study recommendations implementation dates.
- Delaying planned plant turnaround
- Any out of budget expenditure related to process safety that is not approved due to budget constraints

Using Operations & Maintenance Risk Matrix

- If the risk falls in High Risk category, the decision is not acceptable and continued operation of the plants is not allowed unless measures are implemented to lower the risk to medium category.
- If decision is taken to continue to operate in High Risk category, this decision must be officially communicated to and approved by Chief Operating Officer and Managing Director (Occupier of factory).

TATA CHEMICALS LIMITED

41

Using Operations & Maintenance Risk Matrix

- If risk category is medium, additional controls (engineering or administrative) needs to be implemented and risk shall be communicated to and approved by VP-Manufacturing

- If risk category is low, no additional action is required

TATA CHEMICALS LIMITED

42

Using Operations & Maintenance Risk Matrix

- Risks keep dynamically changing
- All the decisions taken through the risk matrix need to be reviewed weekly in the WPM.
- All decisions taken through the Operation and Maintenance Risk Matrix shall be recorded in PROCESS SAFETY EXCEPTION APPROVAL FORM

TATA CHEMICALS LIMITED

43

Process Safety Competency and Knowledge Management

TATA CHEMICALS LIMITED

44

Process Safety Competency and Knowledge Management

TATA

"We do not know what we do not know"

Knowledge and Control of operation

Hazards identification and control

Learning organization

"Organisations do not have memory"

TATA CHEMICALS LIMITED

45

Process Safety Competency

TATA

- Competencies required for process safety is a combination of <u>technical</u> and managerial skills.

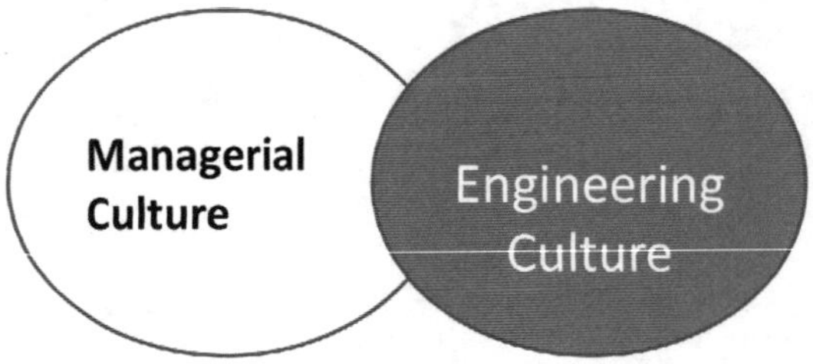

TATA CHEMICALS LIMITED

46

Process Safety Competency

Some of the key process safety competencies are

- understanding SOPs and maintenance procedures,
- key process safety information like MSDS, P&IDs, PFDs, design basis for emergency systems, etc
- use of engineering codes and standards,
- skills for finding root cause for process incidents,
- HAZOP study
- Process of Management of Changes,
- Pre-Design Checklist for Management of Change, etc.

TATA CHEMICALS LIMITED

47

Process Safety Competency Procedure

- Position wise PSM training needs are identified
  - ✓ "Position wise Process Safety Competencies/Training Matrix"
- PSM Training Calendar" –Yearly
- Training & Re-training once every three years
- Pool of trained people in HAZOP
- Pool of trained people in Process Incident Investigation
- Pre-Design Checklist
- Training on the relevant engineering codes & standards – shall be arranged annually by Head – TS & Projects
- All Executives, HOD's and Managers shall be trained annually on the procedure for "Institutionalizing Risk Based approach towards process safety"

TATA CHEMICALS LIMITED

48

Process Safety Knowledge Resources

TATA

BP Process Safety Series

1. Hazards of Water
2. Hazards of Air and Oxygen
3. Hazards of Steam
4. Hazards of Trapped Pressure and Vacuum
5. Confined Space Entry
6. Hazardous Substances in Refineries
7. Hazards of Electricity and Static Electricity
8. Hazards of Nitrogen and Catalyst Handling
9. Control of Work
10. Safe Ups and Downs for Process Units

TATA CHEMICALS LIMITED

49

Process Safety Knowledge Resources

TATA

CSB Videos

AICHE : Safety in Ammonia Plants and Related Facilities Symposium

TATA CHEMICALS LIMITED

50

Other Process Safety KM resources

1. All incidents since commissioning have been captured and uploaded on F drive. The reasons for the incidents are also recorded.
2. Past KM stories related to process incidents shall be retrieved from Knowledge Management portal and shared in weekly training classes
3. Database for RCA reports of Process Safety incidents and
   - one page summary
   - KM portal
4. Learning from other plants / industry

TATA CHEMICALS LIMITED

51

Performance Review Structure – Safety First

Safety takes priority over everything – efficiency, production and maintenance

TATA CHEMICALS LIMITED

THANKS

Acknoledegements:
Dupont Safety ppts
Prism Consultants

TATA CHEMICALS LIMITED

53

Backup Slides

TATA CHEMICALS LIMITED

54

9 Golden Safety Rules

TATA

1. **Permit to Work**
2. **Working at Height**
3. **Energy Isolation**
4. **Confined Space Entry**
5. **Lifting Operations**
6. **Standard Operating Procedures & MOC**
7. **Vehicle Safety**
8. **Ground Disturbance**
9. **Storage of Hazardous Chemicals**

TATA CHEMICALS LIMITED

55

Explanation of terms

TATA

PROBABILITY

- **Frequent :** 1 in 2 years or less:

Based on actual experience and plant incident database

- **Probable:** One in 5 years

Based on actual experience and plant incident database

- **Occasional:** One in 10 years

Based on actual experience and plant incident database

- **Remote:** One in 20 years

Has happened in industry rarely

- **Improbable: One** in 50 years

Not possible to happen in plant lifetime (not happened in other plants also)

TATA CHEMICALS LIMITED

56

Explanation of terms

IMPACT

People:

- Multiple fatalities
- Single fatality or Permanent disability
- Lost Work Day cases (LWC)
- Medical treatment cases or Restricted Work Case (MTC or RWC)
- First aid case (FAC)

TATA CHEMICALS LIMITED

57

Explanation of terms

IMPACT

Assets:

- Extensive damage: Multiple equipment damaged (explosions etc)
- Major damage: Single line equipment damage resulting in plant stoppage or any other damage resulting in continued operation with reduced load
- Local damage: Stoppage of one section of plant
- Minor damage: Plant operations managed with standby equipment
- Slight damage: No effect on production/safety

TATA CHEMICALS LIMITED

58

Explanation of terms

IMPACT

Environment:

- Massive effect: Impacts outside factory area
- Major effect: Impacts within factory area
- Localised effect: Impacts within individual plant possible
- Minor effect: Impacts within section of plant
- Slight effect: Reversible

Explanation of terms

IMPACT

Reputation:

- Worldwide Impact: International media interest
- National Impact: National Media interest
- Regional: Regional Media interest
- Minor impact: Possible change in employee perception of safety
- Slight impact: No significant change

Cranfield UNIVERSITY School of Management

High Reliability Organizing for Process Safety: Principles, Practices and Leadership

CCPS Process Safety Summit
Mumbai, India, 15 Dec 2014

Charles Cowley
Doctoral Researcher, Cranfield University, UK, School of Management
(CCPS Staff Consultant, ex Shell)

 Page 1

Background

Cranfield UNIVERSITY School of Management

Rapid industrialization since the mid-20th century
- **Many serious incidents**:

- Texas City 1947
- Flixborough 1974
- Seveso 1976
- Tenerife 1977
- Chicago DC10 1979
- Three Mile Island 1979
- Mexico City 1984
- **Bhopal 1984**
- **Chernobyl 1986**
- Zeebrugge 1987
- Piper Alpha 1988
- Exxon Valdez 1989

 Page 2

Bhopal remembered…

Cranfield UNIVERSITY School of Management

The MIC Storage tanks E610, 611 and 619 in 2014 (Julia Green www.bhopal.org)

Historical approaches to improvement

Cranfield UNIVERSITY School of Management

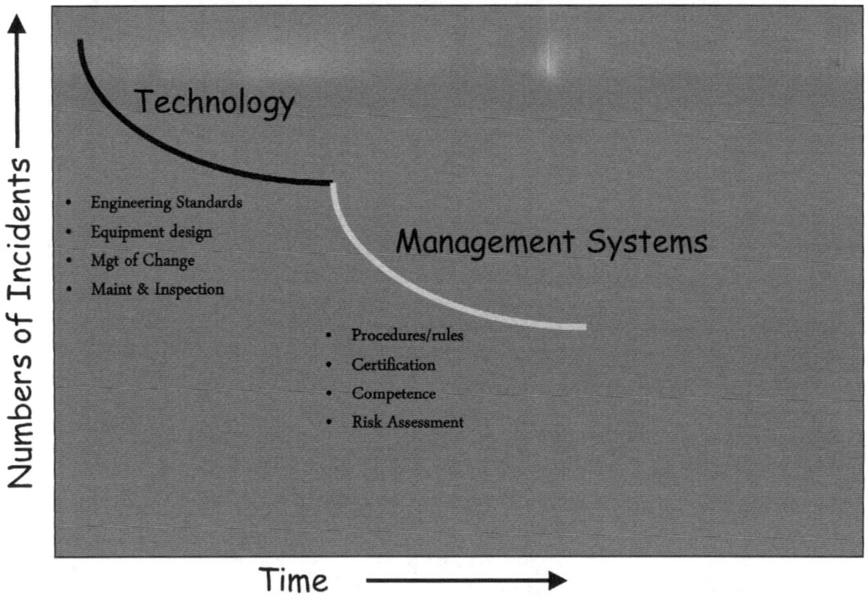

Background to HRO

Cranfield UNIVERSITY School of Management

- **1980s -** cross-disciplinary (psychologists & sociologists) research in the US, on several organizations that, despite operating with **high hazards,** appeared to have **very few accidents**:

Aircraft carriers

Air Traffic Control

Commercial Nuclear Power

- **....Beginnings of 'HRO' theory** (Karlene Roberts, Todd LaPorte, Gene Rochlin)

HRO Theory – some key ideas from 80s research

Cranfield UNIVERSITY School of Management

HROs have active organisational and interpersonal processes that reduce and contain human errors and system failures:

- *"**everyone knows the big picture and constantly communicates that understanding**..."*
- *leaders encourage **strong responses to weak signals** by rewarding reporting of errors and defects and avoiding individual blame;*

HROs are adept at coping with paradoxes: eg:

- ***standardisation AND flexibility**: standardised procedures AND deliberate flexibility to encourage creativity in problem-solving:*
  ...some organisational 'slack' and multi-skilling are needed
- ***trust AND challenge** / respectful cross-checking (good teamwork)*

Prof Karlene Roberts, University of California - Berkeley 1990

More HRO Theory (Karlene Roberts)

Cranfield UNIVERSITY
School of Management

Non-HROs

- **Focus on past success**
- **Underdeveloped cognitive infrastructure**
- **Focus on efficiency**
- **Inefficient learning (episodic)**
- **Lack of diversity (focused conformity)**
- **Information & communications filtering**
- **Reject early warning signs of quality degradations**

EXAMPLE of HROs: Commercial Aviation

Cranfield UNIVERSITY
School of Management

Fatal Accidents – Worldwide Commercial Jet Fleet – 1959 Through 2010

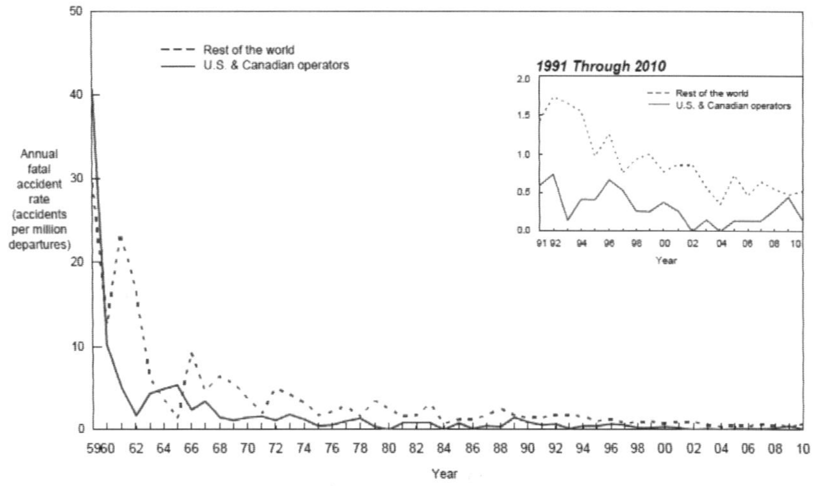

Source: Boeing Commercial Airplanes 2010 STATISTICAL SUMMARY, JUNE 2011

18
2010 STATISTICAL SUMMARY, JUNE 2011

BOEING

Accidents continue in 21st century

Cranfield UNIVERSITY School of Management

Despite improved engineering and management, serious incidents continue…

- Texas City 1947
- Flixborough 1974
- Seveso 1976
- Tenerife 1977
- Chicago DC10 1979
- Three Mile Island 1979
- Mexico City 1984
- **Bhopal 1984**
- **Chernobyl 1986**
- Zeebrugge 1987
- Piper Alpha 1988
- Exxon Valdez 1989

- Columbia Space Shuttle, USA 2003
- Texas City Refinery, USA 2005
- Buncefield, UK 2005
- Guangxi, China 2008
- Jaipur, India 2009
- Yichun, China 2010
- Gulf of Mexico Macondo 2010
- Fukushima, Japan 2011
- Xintai, China 2011
- Lac-Mégantic, Canada 2013
- Visakhapatnam refinery, India 2013
- Soma Manisa, Turkey coal mine 2014
- Kaohsiung, China 2014
- Changsa, China 2014

Background to HRO

Cranfield UNIVERSITY School of Management

- ***Today - renewed interest in HRO in the process industry after the continued series of major incidents in the 2000s***

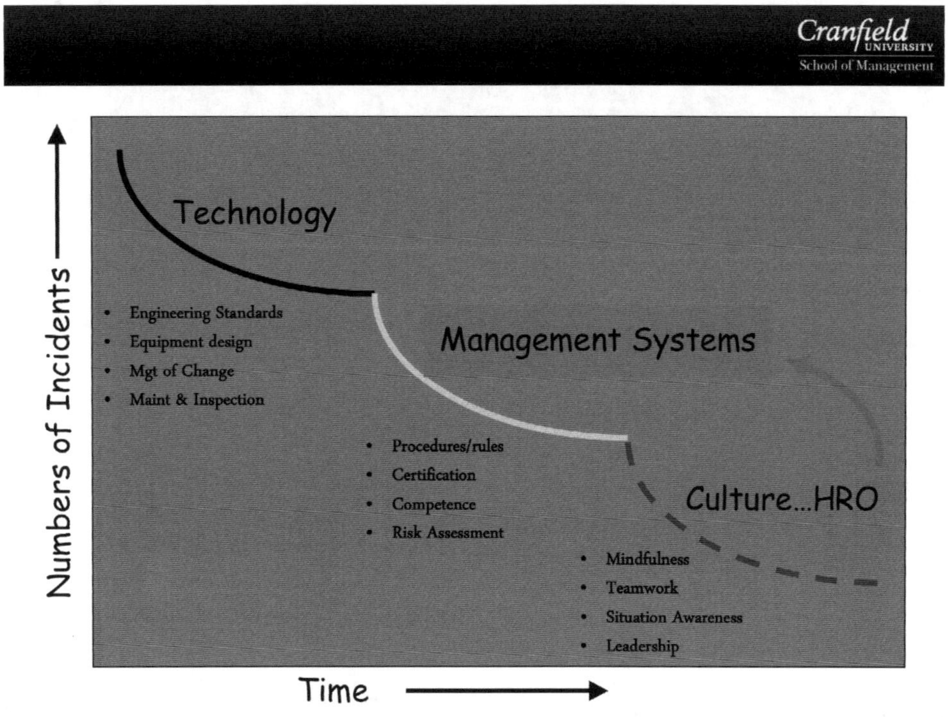
Cranfield UNIVERSITY
School of Management
Technology
Engineering Standards
Equipment design
Mgt of Change
Maint & Inspection
Management Systems
Procedures/rules
Certification
Competence
Risk Assessment
Culture...HRO
Mindfulness
Teamwork
Situation Awareness
Leadership
Numbers of Incidents
Time

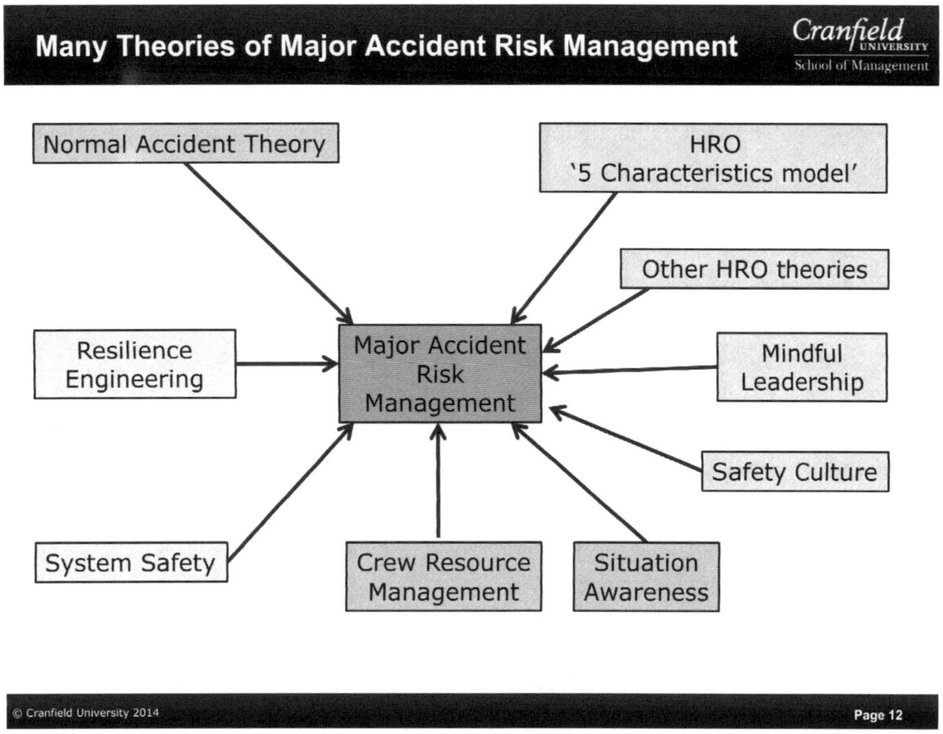
Many Theories of Major Accident Risk Management
Cranfield UNIVERSITY
School of Management
Normal Accident Theory
HRO
'5 Characteristics model'
Other HRO theories
Resilience Engineering
Major Accident Risk Management
Mindful Leadership
Safety Culture
System Safety
Crew Resource Management
Situation Awareness
© Cranfield University 2014
Page 12

HRO Theory – some key ideas... 2001

Cranfield UNIVERSITY School of Management

- **'Mindful sense-making'...**
  - Preoccupation with failure
  - Reluctance to simplify explanations
  - Sensitivity to operations
  - Commitment to resilience
  - Deference to expertise

Weick & Sutcliffe 2001

HRO - 5 Characteristics of Mindfulness (W&S)

Cranfield UNIVERSITY School of Management

- **Preoccupation with failure**
  - 'Chronic unease' .....Reporting culture .....Strong response to weak signals
- **Reluctance to simplify explanations**
  - Avoid loss of critical information AND capacity for analysis, reflection, learning
- **Sensitivity to operations**
  - High situational awareness: control room and boardroom
- **Commitment to resilience**
  - 'a single error must not result in an accident' (T Kletz) ...learning from errors
- **Deference to expertise**
  - Mindful competent decision-making with incomplete info or in unforeseen conditions

HRO - 5 Characteristics of Mindfulness (W&S)

Cranfield UNIVERSITY School of Management

- **Preoccupation with failure**
  - 'Chronic unease' .....Reporting culture .....Strong response to weak signals
- **Reluctance to simplify explanations**
  - Avoid loss of critical information AND capacity for analysis, reflection, learning
- **Sensitivity to operations**
  - **High situational awareness:** control room and boardroom
- **Commitment to resilience**
  - 'a single error must not result in an accident' (T Kletz) ...learning from errors
- **Deference to expertise**
  - Mindful competent decision-making with incomplete info or in unforeseen conditions

 Page 15

Situation Awareness

Cranfield UNIVERSITY School of Management

"In HROs everyone knows the big picture and constantly communicates that understanding..." Karlene Roberts

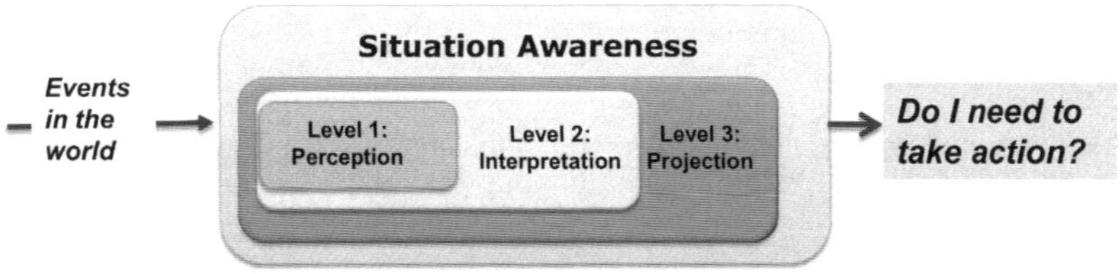

- Level 1: **Perception** of information about what is happening
- Level 2 - **Interpretation** of what the information means for me and my task
- Level 3 - **Projection** of what is likely to happen in the immediate future

Dr. Mica Endsley http://www.satechnologies.com/

Neatly summarised as: What? So what? Now what?

HRO - 5 Characteristics of Mindfulness (W&S)

Cranfield UNIVERSITY School of Management

- **Preoccupation with failure**
  - 'Chronic unease' .....Reporting culture ... **Strong response to weak signals**
- **Reluctance to simplify explanations**
  - Avoid loss of critical information AND capacity for analysis, reflection, learning
- **Sensitivity to operations**
  - High situational awareness: control room and boardroom
- **Commitment to resilience**
  - 'a single error must not result in an accident' (T Kletz) ...learning from errors
- **Deference to expertise**
  - Mindful competent decision-making with incomplete info or in unforeseen conditions

 Page 17

Weak signals – Weick and Sutcliffe

Cranfield UNIVERSITY School of Management

- The key difference between HROs and other organisations...often occurs in the earliest stages, when the unexpected may give off only weak signals of trouble
- The overwhelming tendency is to respond to weak signals with a weak response. **Mindfulness** preserves the capability to see the significant meaning of weak signals and to give strong responses to weak signals
- HROs don't necessarily see discrepancies any more quickly, but when they do spot discrepancies, they understand their meaning more fully and can deal with them more confidently

Weick, K.E. and Sutcliffe, K.M. (2007) Managing the Unexpected. Resilient Performance in an Age of Uncertainty. 2nd Edition John Wiley

Weak Signals: event detection

Cranfield UNIVERSITY School of Management

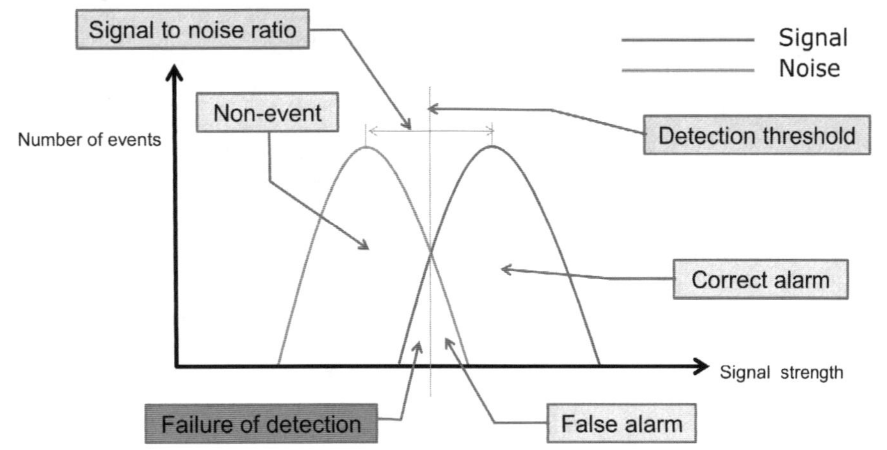

 Page 19

Fast and Slow Thinking... Daniel Kahneman

Cranfield UNIVERSITY School of Management

'Fast' thinking

- Intuitive
- Effortless

- Jumps to conclusions
- Suppresses ambiguity
- Automatic (Always on)

'Slow' thinking

- Deliberate
- Effortful

- Reasoning, Checking
- Sees ambiguity; experiences doubt
- Controlled (Off unless turned on)

Daniel Kahneman, 'Thinking, Fast and Slow' 2011

'Normal operating'
...systems; procedures;
Operational Discipline

'Mindful sense-making'
...

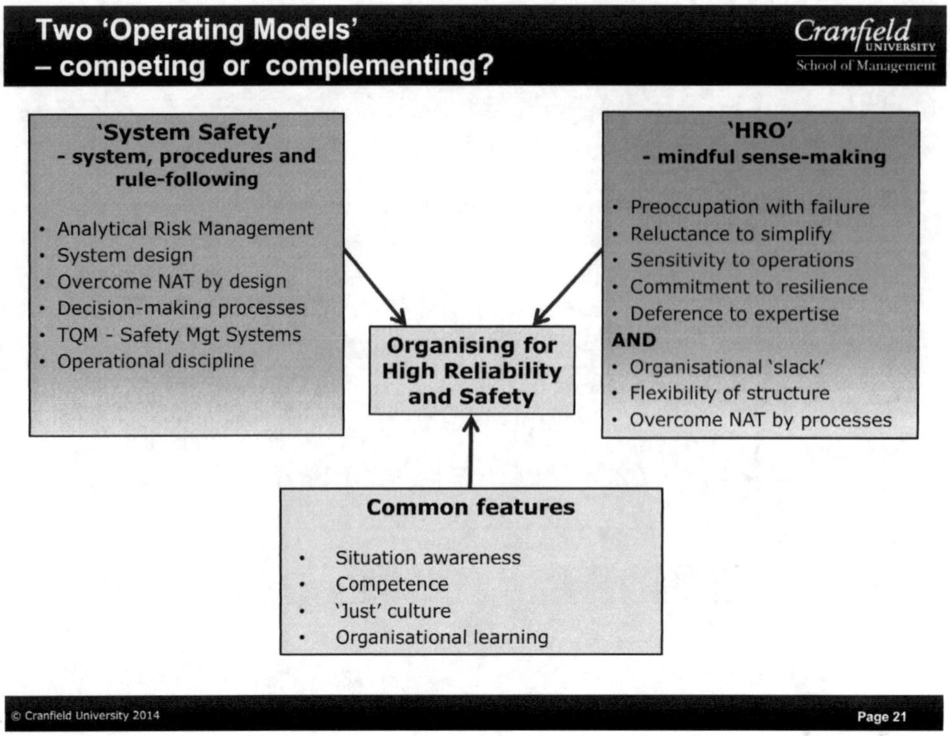
Two 'Operating Models'
– competing or complementing?
Cranfield UNIVERSITY
School of Management
'System Safety'
- system, procedures and rule-following
• Analytical Risk Management
• System design
• Overcome NAT by design
• Decision-making processes
• TQM - Safety Mgt Systems
• Operational discipline
'HRO'
- mindful sense-making
• Preoccupation with failure
• Reluctance to simplify
• Sensitivity to operations
• Commitment to resilience
• Deference to expertise
AND
• Organisational 'slack'
• Flexibility of structure
• Overcome NAT by processes
Organising for High Reliability and Safety
Common features
• Situation awareness
• Competence
• 'Just' culture
• Organisational learning
© Cranfield University 2014
Page 21

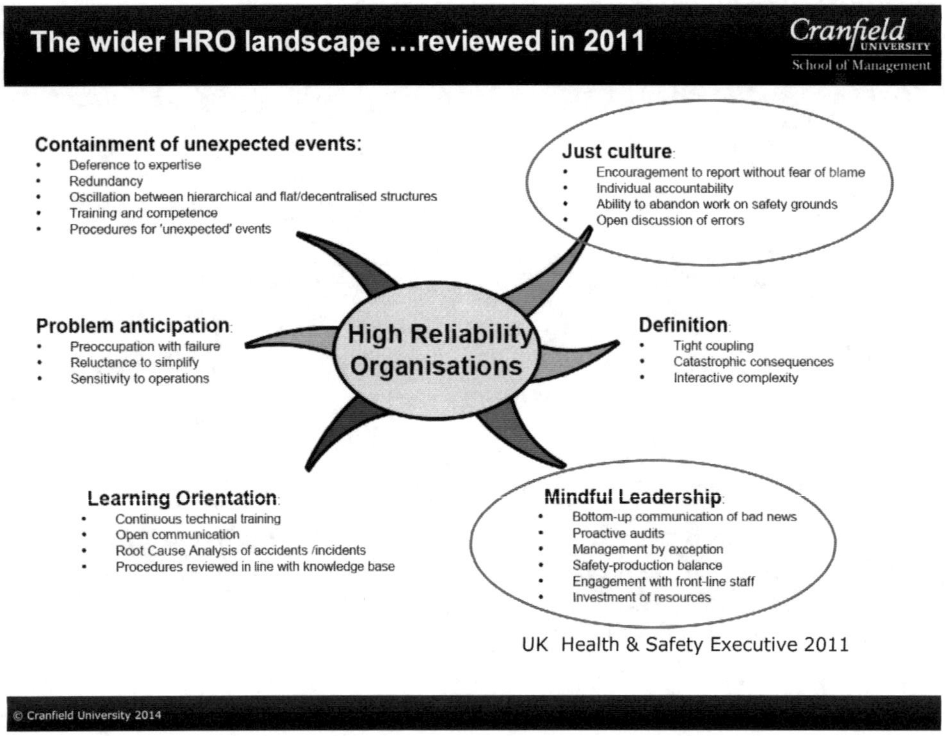
The wider HRO landscape ...reviewed in 2011
Cranfield UNIVERSITY
School of Management
Containment of unexpected events:
• Deference to expertise
• Redundancy
• Oscillation between hierarchical and flat/decentralised structures
• Training and competence
• Procedures for 'unexpected' events
Just culture:
• Encouragement to report without fear of blame
• Individual accountability
• Ability to abandon work on safety grounds
• Open discussion of errors
Problem anticipation:
• Preoccupation with failure
• Reluctance to simplify
• Sensitivity to operations
High Reliability Organisations
Definition:
• Tight coupling
• Catastrophic consequences
• Interactive complexity
Learning Orientation:
• Continuous technical training
• Open communication
• Root Cause Analysis of accidents /incidents
• Procedures reviewed in line with knowledge base
Mindful Leadership:
• Bottom-up communication of bad news
• Proactive audits
• Management by exception
• Safety-production balance
• Engagement with front-line staff
• Investment of resources
UK Health & Safety Executive 2011
© Cranfield University 2014

'Leadership for Process Safety'

Cranfield UNIVERSITY
School of Management

- **OECD Guidance**
  - Guidance for Senior Leaders in High Hazard Industries
- **OGP 452** –Shaping Safety Culture through Safety Leadership
- **CCPS**
  - Inspiring Process Safety Leadership
    - The Executive Role
- **IChemE / EPSC**
  - Process Safety Leadership
    for Senior Executives (PSL)
- **UK HSE**
  - Leadership for the major hazard industries
  - Process Safety Leadership Group 'Principles'
- ....other

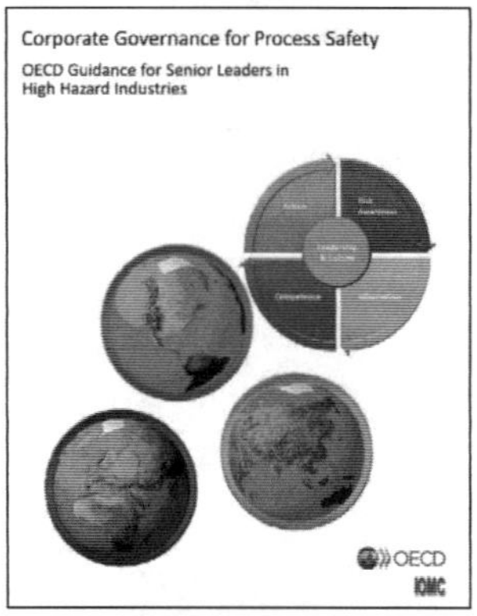

'Mindful Leadership' – some key components?

Cranfield UNIVERSITY
School of Management

- Chronic Unease
- 'Just' culture
- Engaging – especially with front line operations
- Sense-making and Integrating
- Committing to provide adequate resources
- Challenging and supporting people...

Is this how Leadership influences HRO development? (This is CURRENT RESEARCH...)

Cranfield UNIVERSITY School of Management

Context eg:
- Organization's type of activity: Technological Sociological
- Hazard Consequences
- Regulatory Political Litigational
- Other...

HRO 'Principles'
- **Preoccupation with failure**
- **Reluctance to simplify**
- **Sensitivity to operations**
- **Commitment to resilience**
- **Deference to expertise**

and...
- Redundancy
- Flexibility of structure
- Competence
- Organisational learning
- System Safety

Leadership roles
eg:
- Chronic unease
- 'Just' culture
- Engaging
- Sense-making & Integrating
- Allocating resources
- Challenging & Supporting

HRO Components:
Processes, Structures, Systems, Practices ...eg:
- Decision processes
- Centralised organisation
- Use of Leading Indicators
- Operational discipline
- Error Management /CRM
- Situation Awareness

© Cranfield University 2014 Page 25

EXAMPLE: Gulf of Mexico oil well blowout 2010

Cranfield UNIVERSITY School of Management

© Cranfield University 2014 26

EXAMPLE: Gulf of Mexico oil well blowout 2010

Cranfield UNIVERSITY School of Management

Context eg:
- Organization's type of activity: Technological Sociological
- Hazard Consequences
- Regulatory Political Litigational
- Other...

HRO 'Principles'
- **Preoccupation with failure**
- **Reluctance to simplify**
- **Sensitivity to operations**
- **Commitment to resilience**
- **Deference to expertise**

and...
- Redundancy
- Flexibility of structure
- Competence
- Organisational learning
- System Safety

Leadership roles

eg:
- Chronic unease
- 'Just' culture
- Engaging
- Sense-making & Integrating
- Adequate resourcing
- Challenging & Supporting

HRO Components:

Processes, Structures, Systems, Practices ...eg:
- Decision processes
- Centralised organisation
- Use of Leading Indicators
- Operational discipline
- Error Management /CRM
- Situation Awareness

 Page 27

EXAMPLE: QF32 – A380 Engine Loss 4 Nov 2010 ...BUT...disaster averted

Cranfield UNIVERSITY School of Management

Figure 16: No. 2 engine showing fire damage to the lower left engine nacelle

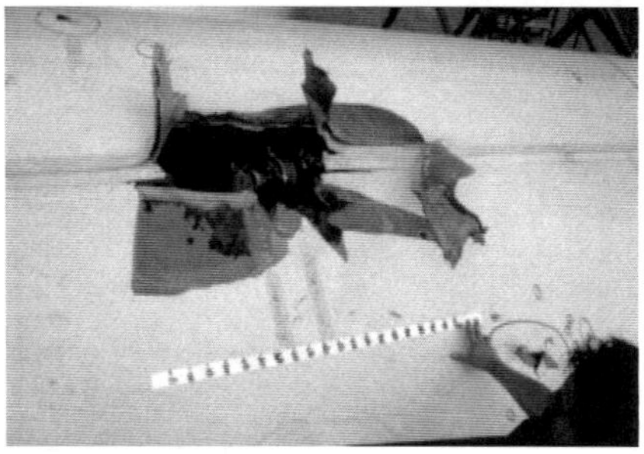

 Page 28

EXAMPLE: QF32 – A380 Engine Loss 4 Nov 2010 ...BUT...disaster averted

Cranfield UNIVERSITY School of Management

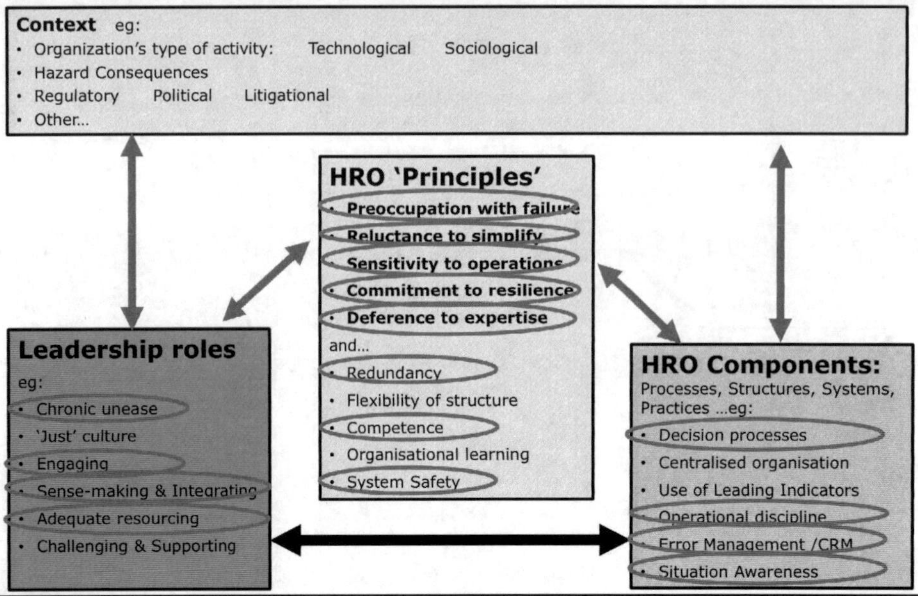

 Page 29

Focus of current research

Cranfield UNIVERSITY School of Management

- To identify **leadership roles and practices**
- ...that are effective in influencing the development of Highly Reliable Organisations
- Contact **charles.cowley@cranfield.ac.uk**

 Page 30

KEY TAKE-AWAYS

Cranfield UNIVERSITY
School of Management

- As well as Systems, Procedures and 'Operating Discipline' we need '**Mindful Sense-making**' to achieve **Situational Awareness**
- **Principles of Mindfulness** include:
  - Preoccupation with failure
  - Reluctance to simplify explanations
  - Sensitivity to operations
  - Commitment to resilience
  - Deference to expertise
- '**Mindful Leadership**' practices that contribute may include:
  - Chronic unease
  - 'Just' culture
  - Engaging – especially with front line operations
  - Integrating and Sense-making
  - Adequate resourcing
  - Challenging and Supporting

 Page 31

Cranfield UNIVERSITY
School of Management

Thank you…!

Questions ?

 Page 32

BACK-UP

Cranfield UNIVERSITY School of Management

- **EXTRA SLIDES for further reading or questions…**

Situation Awareness

Cranfield UNIVERSITY School of Management

It is very easy to LOSE (or never gain) high Situation Awareness -

- Poor design of operator / system interface
- Distraction
- Fatigue
- Stress
- Poor communication within team
- Low motivation

- Poor information about the plant, the operating environment...
- Lack of knowledge about what is going on down in the shop floor...
- Lack of information about process disruptions, near-incidents ....
- Poor auditing...

© Cranfield University 2014

Situation Awareness

Cranfield UNIVERSITY School of Management

High Situation Awareness takes effort

- Ergonomic design of operator / system interface
- Avoid distractions (alarm management; sterile area... quiet room)
- Fatigue management (shift rostering... / breaks...
- Stress monitoring / support / coping strategies...
- Communication within team (Crew Resource Management... Radio discipline)
- Team / shift motivation (supervisor training...)

- Routine recurrent training; use of simulators...
- Pre Job Briefings... Frequent team / shift meetings
- Rapid sharing of information from plant... Frequent management site visits
- Learning From Incidents: process disruptions, near-incidents ....
- High value auditing...

 Page 35

Cranfield UNIVERSITY School of Management

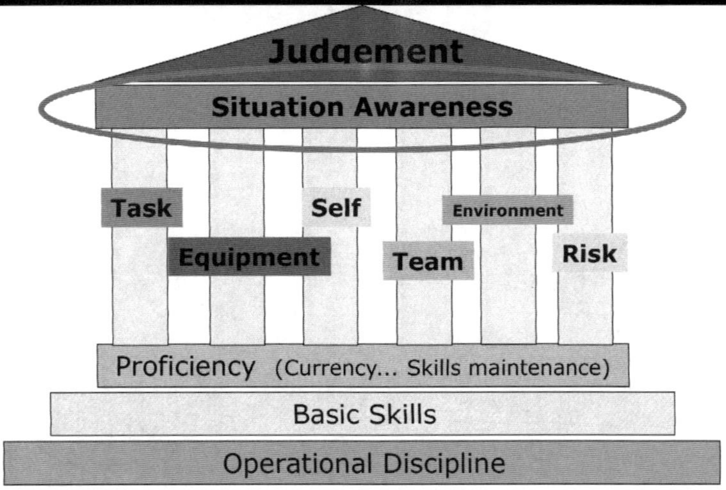

Tony Kern's Airmanship model, applied to process plant

 Page 36

Culture Ladder

Cranfield UNIVERSITY
School of Management

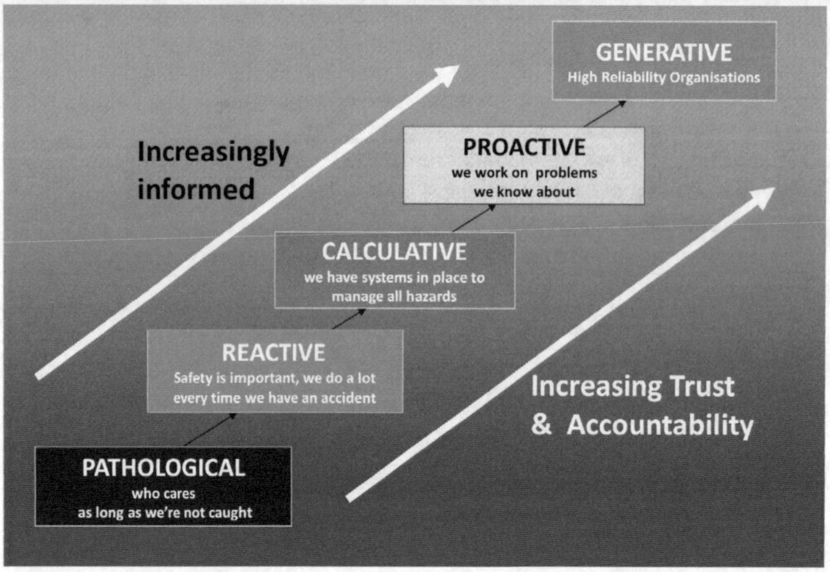

Cranfield UNIVERSITY
School of Management

Available from Energy Institute Hearts and Minds - Home

Use of Leading and Lagging Indicators

Cranfield University School of Management

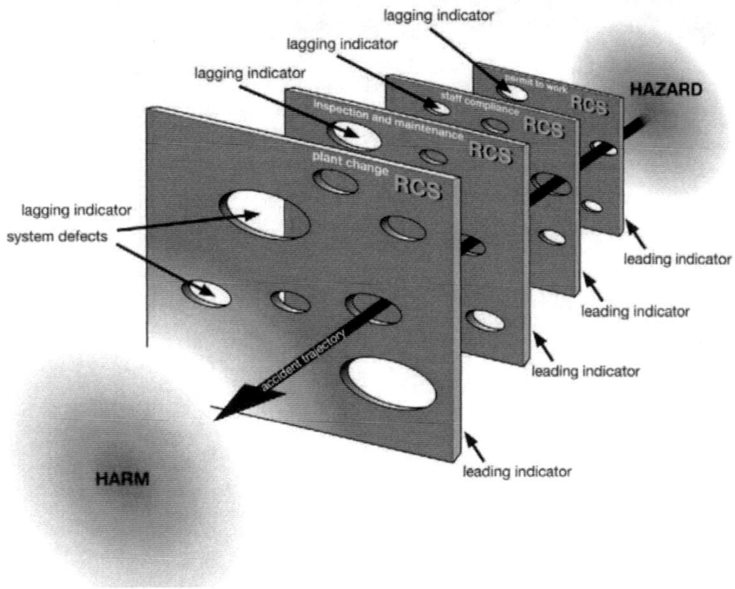

Further reading

Cranfield University School of Management

CCPS Risk Based Process Safety Management - 2007

CCPS Conduct of Operations and Operational Discipline - 2011

UK H&S Exec RR899 – HRO Literature Survey - Nov 2011

Leadership and HROs - why banks fail Young, B. 2012

* OGP 452 Safety culture and leadership - **Oct 2013**
* OGP 460 Cognitive issues in process safety - **Jul 2012**
* OGP 510 Operating Management Systems - **Jun 2014**
* OGP 511 OMS in practice - **Jun 2014**

Energy Institute 'High level framework for Process Safety' Dec 2010

http://high-reliability.org/pages/High-Reliability-Organizations

An AIChE Technology Alliance

CCPS

Center for Chemical Process Safety

TÜV NORD

Vibrant Management Framework for Enhancing Process Safety

By S.Raghu.
Vice President –Engineering Services
South East Asia

December 2014 –Mumbai

TÜV NORD

- Disclaimer:
- **This presentation is based on my views and experiences and does not represent the view of my current employer.**

THIS PRESENTATION IS BASED ON PUBLICATION /STUDY IN WHICH INSPECTION /AUDIT REPORTS FROM PERIODICAL INSPECTIONS OF PLANTS HAVE BEEN USED

Introduction

▪Purpose of Presentation.

▪Provide an update on the Development of the Level Process Safety

▪Vibrant Management Framework

- •The need.
- •The outline of the Framework
- •Implementation through Good Practices to enhance Process safety for Vibrant Management system

and Further development of the Framework using Excellence Model

An Outline on Vibrant Management System

- **In Vibrant Management Systems:**
- Envisage a very high Participation of all employees to develop a clear understanding of the expectations of senior management and those expectations should be documented and shared.
- **There is a structure that clearly defines and documents expectations for all systems that affect process safety.**
- **The management systems promote inherently safer design principles and the principles of RBPS.**
- **The word vibrant is used to describe the management system because the system needs to be flexible and adaptable to meet ever changing needs and it needs to work well with facilities of all sizes in the organization.**
- **Companies with great process safety performance do not impose the most complex systems on their facilities; instead, they require systems that are fit-for-purpose.**
- Risk Based Process Safety (RBPS) provides many alternatives for implementation of process safety element management systems and is a useful tool when companies are developing
- or refining their fit-for-purpose systems

Vibrant management system 2015/1/13 4

Process safety management in context

TÜV NORD

The challenge for the process industry is often clearly set out in three statements by chief executives as:

To ensure we don't hurt anyone;
To ensure we don't harm the environment,
and to achieve a certain level of return on capital employed (ROCE).

Typically, organizations are very clear on what they need to do to deliver ROCE - maximize income and optimize expenditure - however, without a clear understanding of what is required to meet health, safety and environment (HS&E) objectives including PSM, the approach and costs of managing them will lack control. Furthermore, there may be little indication or warning if these activities are being ineffective until an incident occurs.

Key questions

TÜV NORD

We have all seen the typical banner statements 'zero harm', 'flawless operation', 'target zero', 'incident free', nobody gets hurt'; but the two key questions for executives and managers at all levels are:

1. How will we assure the integrity of the operation?

2. How will we <u>know</u> we are doing it?

All too often the first two words used to answer these questions are "I think...."; in reality this means "I don't know"! Recent events have shown that such answers are no longer acceptable and that, from top to bottom, organizations need to be able to answer these two key questions with absolute confidence.

Background

TÜV NORD

- Process Safety Committees have identified benefit in promoting a common high-level framework for Process Safety Management across the energy industry sectors and Various questions arising :
- Development Approach
- Workshops with participants representing a cross section of the Energy Sector including the HSE.
- Challenge & Validate need for a Process Safety
- Management Framework.
- Identify what we need from a PSM Framework
- Define the scope & detail of Framework
- Subsequent wider interest group review

Background

TÜV NORD

- Implementing Vibrant Management system is more prominent -Given the Clear Need after realizing :
  - **Past Disasters**
  - **Current Year Disasters**
  - **Perceived Weakness in PSM Program**

The need

▪There is a clear Global drive :

-to support the need for a

▪**High Level Framework for Vibrant Process Safety Management.**

▪–Provide a vision for PSM

▪–Define a model for and raise profile of PSM within a company

▪–Define what needs to be in place to maintain "Licence to Operate"

▪–Help to define roles, responsibilities

▪& accountabilities for PSM

▪–Easy to understand how to achieve suitable PSM

▪–Provide a basis for benchmarking

Vibrant Management System Create Advantages

- **Integrity of operation –Fit for Purpose**
- **Maximize Income**
- **Minimize Costs**
- **Effective Capital Investment**
- **Effective People &Organization**

To Build a Vibrant Management System, then what? First... TÜV NORD

- Form a Team in Your Company, i.e..
  - Process Engineers
  - Operators
  - Safety
  - Maintenance
  - Management
  - Consultants

Remember...You Can't Do it Alone!

Now that we are required to build Vibrant Mangement system , then what? Second... TÜV NORD

- Form a Plan, Determine:
  - Responsibilities
  - Duties
  - Reporting
  - Document Control
  - Progress Reports
  - Tracking Changes

Now that we are required to build Vibrant management System , then what? Third...

TÜV NORD

- Review Scope :
  - Process Safety Information
  - Equipment in the Process
  - Process Hazard Analysis
  - Operating Procedures
  - Employee Training
  - Training Contractors
  - Pre-Startup Safety Review
  - Mechanical Integrity of Equipment
  - Management of Change
  - Incident Investigation
  - Emergency Preparedness
  - Compliance Audits
  - Trade Secrets

Now that we are required to Build Vibrant management system, then what? ...Let's Begin

TÜV NORD

- we will Begin to Explore key Element of the Standard and What is Required to Comply

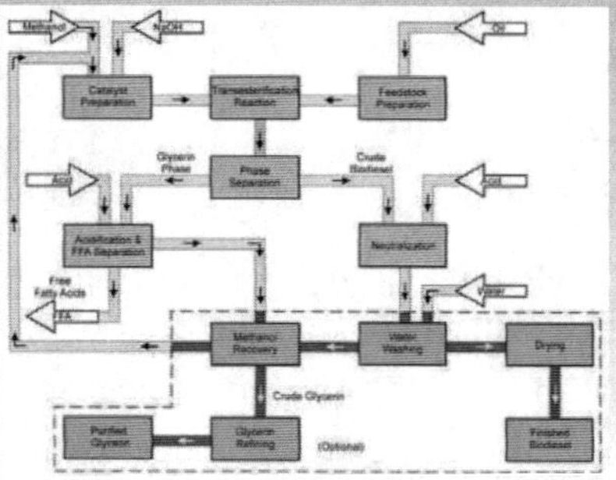

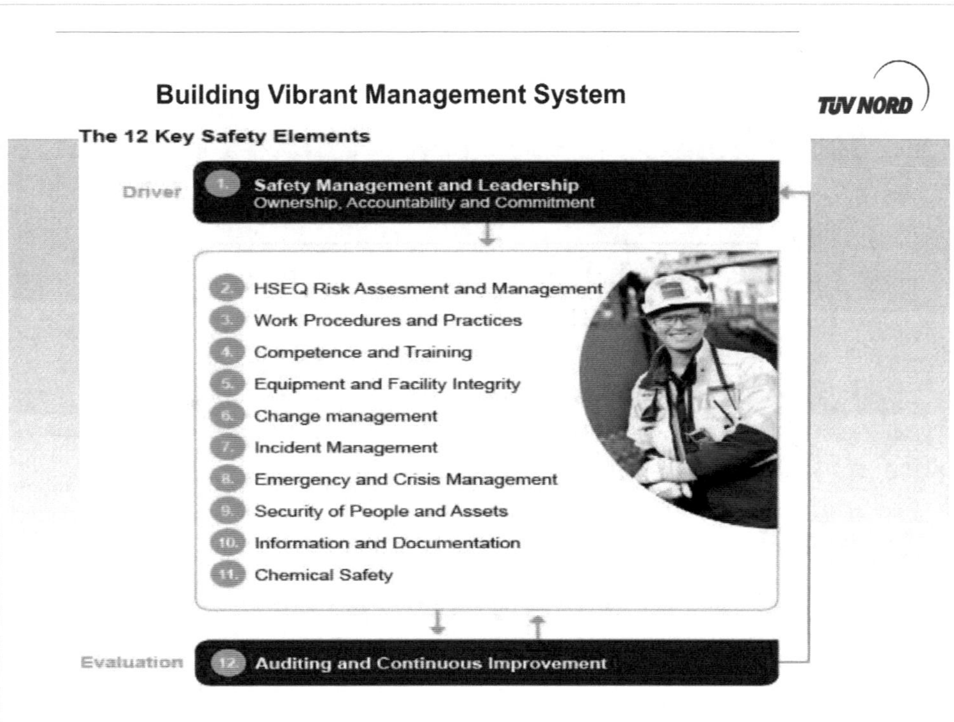

Building Vibrant Management System
TÜV NORD
The 12 Key Safety Elements
Driver
1. Safety Management and Leadership
Ownership, Accountability and Commitment
2. HSEQ Risk Assesment and Management
3. Work Procedures and Practices
4. Competence and Training
5. Equipment and Facility Integrity
6. Change management
7. Incident Management
8. Emergency and Crisis Management
9. Security of People and Assets
10. Information and Documentation
11. Chemical Safety
Evaluation
12. Auditing and Continuous Improvement

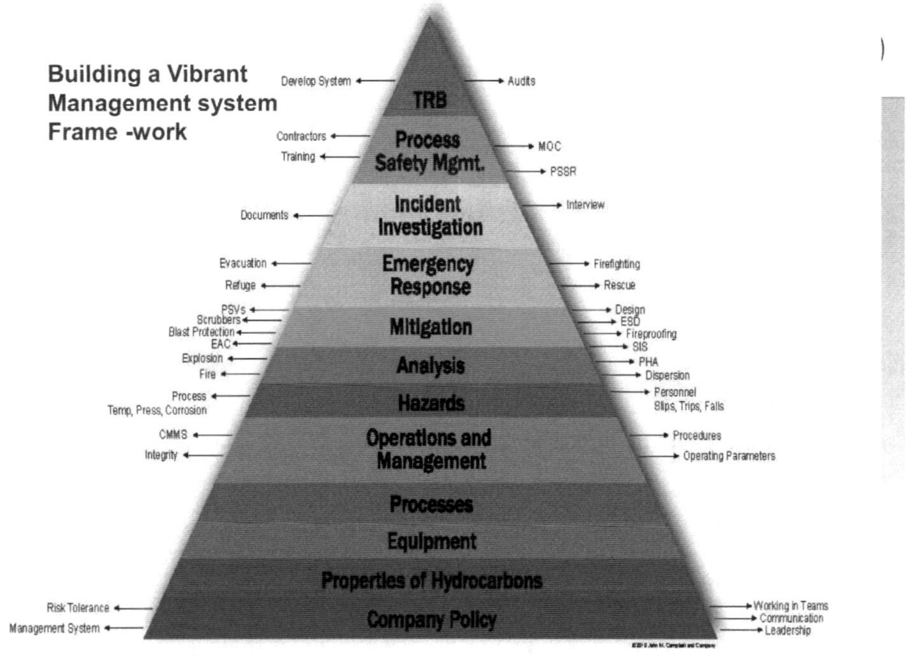

Building a Vibrant Management system Frame -work
Develop System
Audits
TRB
Contractors
Training
Process Safety Mgmt.
MOC
PSSR
Documents
Incident Investigation
Interview
Evacuation
Refuge
Emergency Response
Firefighting
Rescue
PSVs
Scrubbers
Blast Protection
EAC
Mitigation
Design
ESD
Fireproofing
SIS
Explosion
Fire
Analysis
PHA
Dispersion
Process
Temp, Press, Corrosion
Hazards
Personnel
Slips, Trips, Falls
CMMS
Integrity
Operations and Management
Procedures
Operating Parameters
Processes
Equipment
Properties of Hydrocarbons
Risk Tolerance
Management System
Company Policy
Working in Teams
Communication
Leadership

To monitor and measure Process safety

How can we measure process safety performance?
A person first needs to get a true understanding of the words: ***process, safety, management.***
The first word is **Process**. PSM is concerned with process issues such as fires, explosions and the release of toxic gases caused by process-oriented issues such as corrosion, chemical reaction and the inadvertent mixing of hazardous chemicals .
The second word in the term PSM is **Safety**. The initial driving force for most PSM programs was the need to meet a safety regulation and to reduce safety incidents related to process upsets and hazardous material release.

To monitor and measure

The third word is **Management**. A manager is taken to be anyone who has some degree of control over the process, including operators, engineers and maintenance workers. The word management also means that PSM is not just about equipment and instrumentation, but also covers issues as Employee Participation, Operating Procedures and Management of Change.

Monitor

PSM is a systematic approach to major accident hazard management. It is a part of safety management, which focuses on the concerns of major hazards impacting safety, environmental damage and business losses. Oil & gas companies are significant users of PSM methods, particularly where there are hazardous processes or large inventories of flammable or toxic materials.

Vibrant Management -Monitoring through Indicators

"Lagging" Metrics – a retrospective set of metrics that are based on incidents that meet the threshold of severity that should be reported as part of the industry-wide process safety metric.

"Leading" Metrics – a forward looking set of metrics which indicate the performance of the key work processes, operating discipline, or layers of protection that prevent incidents.

"Near Miss" and other internal Lagging Metrics – the description of less severe incidents (i.e., below the threshold for inclusion in the industry lagging metric), or unsafe conditions which activated one or more layers of protection. Although these events are actual events (i.e., a "lagging" metric), they are generally considered to be a good indicator of conditions that could ultimately lead to a more severe incident.

Vibrant Management System – Further Recommendation

TÜV NORD

It is strongly recommended that all companies incorporate each of these three types of metrics into their internal process safety management system.
Such method of performance measurement of process safety was one of the recommendations by the national commission that investigated the Texas City incident: "BP should develop, implement, maintain, and periodically update an integrated set of leading and lagging performance indicators [or more effectively monitoring the process safety performance of the U.S. refineries by BP's refining line management, executive management (including the Group Chief Executive), and Board of Directors. In addition, BP should work with the U.S. Chemical Safety and Hazard Investigation Board and with industry, labor organizations, other governmental agencies, and other organizations to develop a consensus set of leading and lagging indicators for process safety performance for use in the refining and chemical processing industries."

Vibrant Management to include :

TÜV NORD

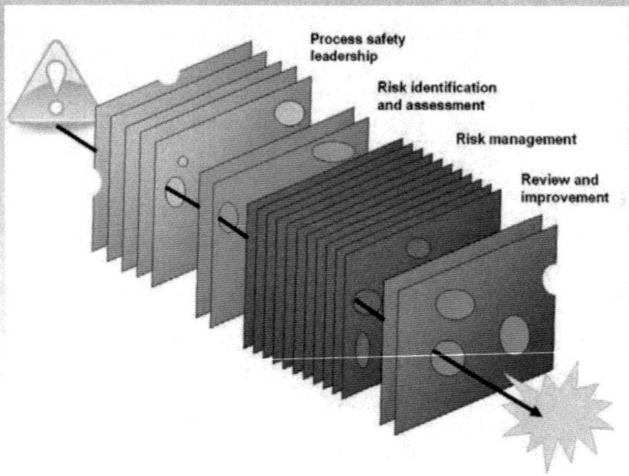

Vibrant management to include Managing: Rule Breaking..

TÜV NORD

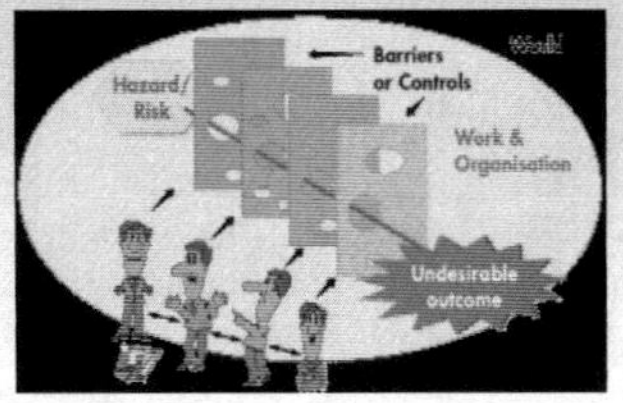

- **Managing Rule Breaking**
- Rules and procedures form one of the major barriers between hazards and unwanted events. Failure to follow these established procedures removes one or more of these barriers. This can be due to human error, or the rules can be broken intentionally. In combination with a single error or mechanical failure, violations can lead to disaster (Swiss cheese model as illustrated next). The Managing Rule Breaking tool will help you to understand why people break rules intentionally and how to manage and change this.

Vibrant management to include Managing: Rule Breaking..

TÜV NORD

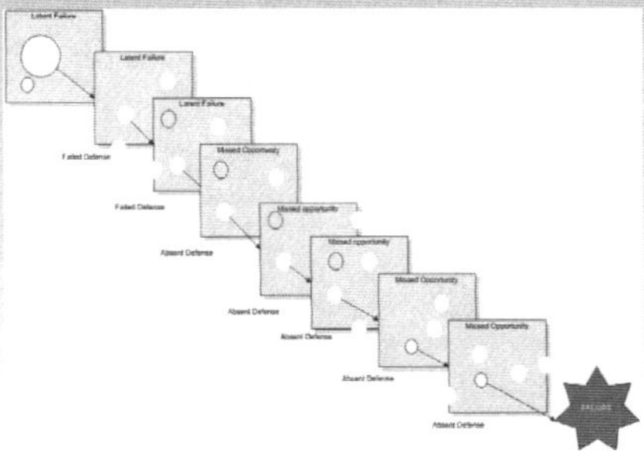

- The majority of people break rules and cut corners for what they see as good reasons - often to get the job done. Managing Rule Breaking explains that most people can be classified as either a Sheep or a Wolf and illustrates how to get both to follow procedures. Sheep are guardians of high standards who dislike having to break the rules. Wolves have fewer problems with violation, feel in control of situations and exploit opportunities when they arise.

PSM

Hierarchy of PSM Elements

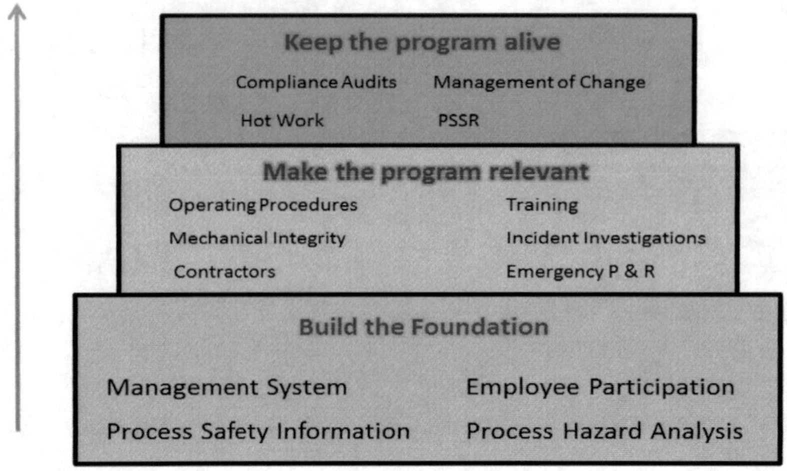

Vibrant Management System

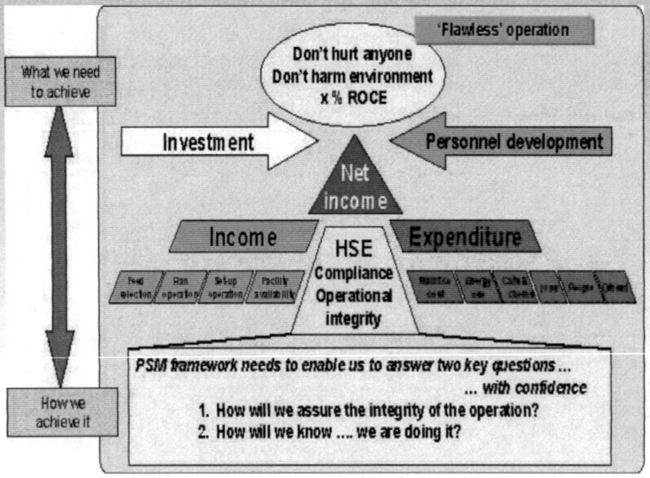

Vibrant Management

TÜV NORD

Key resources - PSM defined and managed in 3 steps
PSM involves managing a number of technical, managerial and human factors activities which, if not managed effectively, may lead to a major incident in future. Three key resources to help industry identify and manage these activities:

1) High level Vibrant framework for process safety management ('PSM framework') - a comprehensive process safety management framework which captures industry good practice in PSM and provides the energy industry with a consistent and effective approach to answer the two key questions with confidence.

2) **Follow an updated** Guidance on meeting expectations / process safety management framework ('PSM guidelines') - comprehensive guidelines to manage process safety for each Element of the *PSM framework*.

3) Process Safety Survey - an in depth survey based on the *PSM framework*, allowing companies to assess their process safety arrangements and benchmark performance against industry.

Building Vibrant Management system using Good Practices..

TÜV NORD

Good Practice for
Enhancing Process Safety

Vibrant mangement system | 2015/1/13 | 34

Sharing Good Practices for vibrant Management –Case Studies

TÜV NORD

- *Awareness of regulatory requirements*
- *Management and personnel commitment*
- *Risk assessment and decision-making*
- *Technical implementation and functionality*
- *Instructions for and assessment of operations*
- *Competence and training*
- *Managing emergencies and deviations*

1.Awareness of regulatory requirements

TÜV NORD

Study Assessment

- How well has the establishment recognized its legal obligations? How are the changes to these obligations monitored?
- For a company to fulfill its statutory obligations, it must be familiar with the legislation related to its operations. It is essential for good compliance with regulatory requirements that the legislation related to operations is monitored regularly, and that this task is carried out by appointed members of staff

Examples of Good Practice

- **Use of press release/CCPS Bulletin to monitor legislation.**
- **-Use of a regulatory information service providing guidance on how the law can be applied to a company's operations.**
- **Registering all comments by authorities and the resulting actions in one system to facilitate access to up-to-date information.**

2.Management and personnel commitment

TÜV NORD

Study Assessment

Are management and staff committed to safety issues? What methods does the management use to maintain, assess and improve the level of safety?

When management and personnel are committed to safety, the latter is seen as an integrated part of the organization's performance as well as an important issue throughout the organization. In a committed culture, safety is regarded as important both at work and during leisure time.
This commitment is enhanced if a company has clear and concrete safety targets that have been embedded in the everyday operations. Targets should be monitored actively to ensure that they are **realized.**

Examples of Good Practice

- An essential factor in gaining personnel commitment is the degree of interest that the management demonstrates towards safety. One way for management to advance safety is to take part in safety tours.
- Communication also plays a significant role in committing personnel to safety. It has to be open and information should be readily accessible to everyone. It is important for the personnel to be able to monitor safety indicators and the progress of actions and initiatives related to development. Everyone should have the opportunity to influence safety. In addition to in-house staff, external service providers should also be taken into account in safety issues

Vibrant mangement system 2015/1/13 37

Examples of good practice

TÜV NORD

Safety communication –More focus on Process safety

Safety issues are the first item on the agenda in all meetings.
Safety meetings (e.g. short meetings on current safety issues) are held regularly.
Safety weeks are organized, including process safety and health-related training.
Awareness of safety-related issues is raised with the help of safety themes, changing monthly or annually.
A safety video is produced, introducing the chemicals in use at the plant, their attendant risks and the action that needs to be taken in the event of an emergency.
Practical comparisons are made on the way in which safety issues are organized between departments, branches and companies within the same sector.
Information is communicated about previous accidents with the help of accident report cards, which contain a summary of the accident with key points illustrated. The cards are accessible to everyone and handled in safety meetings, for example. The cards are also used when planning work, as well as in risk assessment and training.
Safety issues and instructions are communicated via info screens.
Safety issues are raised via paper leaflets distributed on canteen tables, for example.
A safety notice board is maintained with information on safety targets, indicators and previous accidents.
Shift handover meetings are held, during which
any deviations occurring during the previous shift are discussed.
Regularly updated bulletins on current safety issues
(e.g. use of reflectors, international accidents) appear on the intranet landing page or similar internal communication channels.
Operations supervisor meetings are held.

Vibrant Management System-Process Safety 2015/1/13 38

Examples of good practice

TÜV NORD

Employee participation

A development group for safety issues is established, consisting of employees only. A budget is allocated for the group to carry out improvements with immediate effect.

Local chemicals supervisors are appointed in addition to the mandatory operations supervisors.

Observation tours are conducted for new employees.

Safety programmes are put in place for each department.

External employees/visitors

Safety training is implemented for external employees, after which each participant is issued with a picture ID card.

External employees are required to take an exam on orientation material and/or to acknowledge that they have familiarised themselves with it against a signature.

Safety aspects are taken into account when making service agreements.

A safety guide for visitors is launched.

Vibrant mangement system 2015/1/13 39

Examples of good practice

TÜV NORD

Monitoring and documentation

Regular observation tours are conducted by different levels of the organisation.

Various form templates are created for reporting different deviations (e.g. near-misses, accidents, environmental damage, leisure-time accidents).

Service providers are taken into account in accident statistics.

Leading safety indicators are used and developed.

Process safety indicators are developed.

Deviations are entered into and monitored by an ICT system, including information on the processing phase of the deviation, the person in charge and the required corrective/preventive actions. Persons in charge are notified if the actions are not registered as completed by an agreed date.

Rewards

The reporting of near-misses is encouraged (e.g. by handing out meal vouchers).

Staff are rewarded for an accident-free period (e.g. with coffee and cake).

The realisation of safety targets is linked to productivity bonuses.

An 'Employee of the year' award is introduced for actively promoting safety.

Vibrant management system 2015/1/13 40

3.Risk assessment and decision-making

TÜV NORD

Study Assessment

- How has the company assessed the chemical and process risks related to its operations? How are safety issues taken into account in decision-making?
-
- Identifying dangers and risks should be a proactive process, integral to the company's everyday operations. Employees are encouraged to actively identify risks and report any observations. Good practice exists when risk assessments are updated regularly, and immediately when necessary, for example in connection with changes. It is also recommended that several different methods are applied in risk assessment, with the most suitable method being selected depending on the purpose. Risk assessment results must be taken into account in decision-making. Good risk assessment covers all plant operations and is also expected of service providers.

Examples of good practice:

- Information on deviations is distributed between different sites. Based on this information, other sites consider whether a similar situation could occur at their site and whether they are prepared.
- A schedule/watch list for risk assessments is created to indicate when different risk assessments have been carried out.
- Unusual weather conditions are also taken into account in risk assessment.
- Employee interviews are exploited to the full when performing risk assessments.
- Risk assessments are carried out in a project group consisting of representatives from planning, production, maintenance, and management.
- Risk assessments are performed on individual tasks and/or pieces of machinery/equipment.
- A risk assessment is always carried out (e.g with the help of a checklist) before starting a task.
- Risk assessments are reviewed after every deviation.
- Instructions are compiled on how to implement changes. A risk survey is carried out before making any changes.
- In addition to work safety, ENSURE process safety is also taken into account in risk assessments.

Vibrant management | 2015/1/13 | 41

4.Technical implementation and functionality

TÜV NORD

Study Assessment

- How is the functionality of appliances and equipment ensured at a company? How is their reliability monitored and inspected?
- When maintaining the functionality of a plant or piece of equipment, it is important that the plant has an up-to-date maintenance plan and that its implementation is documented. Maintenance should be preventive. Safety requirements must also be taken into account at the equipment acquisition and planning stages. It is also important to utilise any results from maintenance and inspections in order to improve operations

Examples of good practice:

- A maintenance programme is created to help perform preventive maintenance.
- Equipment is classified according to importance and operations so that safety-critical appliances and equipment can be identified. Maintenance and inspections are carried out according to the classification.
- Safety systems and automation are tested regularly.
- Weekly meetings are held in relation to the maintenance system. These assist communication between the plant and those responsible for maintenance.

Vibrant management system | 2015/1/13 | 42

5.Instructions for and assessment of operations

TÜV NORD

Study Assessment:

- What instructions does the company provide for the safe operation of the plant? How are emergencies and deviations taken into account in the instructions?
- Safety can only be advanced if instructions for the safe performance of work tasks are in place. Likewise, everyone working on the company premises must be introduced to safety procedures. It is crucial that agreed procedures and instructions are followed and any deviations are actively addressed. In order to maintain safety, it is important that procedures and instructions are reviewed and updated on a regular basis.

Examples of good practice:

- Safety-critical equipment is identified and work instructions and maintenance in relation to this equipment are defined.
- Process safety is acknowledged as a separate entity.
- Double locking is used during alteration work
- and maintenance. For example, the safety coupling can be locked and the key put in a drawer which is locked by the maintenance staff with their own keys. Once the work is done, the maintenance staff remove their lock.

It is crucial that agreed procedures and instructions are followed.

Vibrant Management system 2015/1/13 43

Examples of good practice

TÜV NORD

The key cannot be taken out until all the locks have been removed. This ensures that the process is not started accidentally during alteration or maintenance work.

Employees are consulted when compiling work instructions in order to verify whether the work can be carried out in the way described.

Work instructions are issued to employees against a signature.

A list of questions is attached to the work instructions to help employees check whether they have understood everything correctly.

Employees are requested to explain what they are doing and why. This helps ensure that the employee is familiar with the instructions and understands why it is important to perform the work in accordance with them.

In addition to the safety datasheet, a separate first-aid guide is compiled with regard to the most dangerous chemicals.

Audits are organised during which employees can observe each other's working methods.

Instructions are handed to service providers, such as drivers at the gate, requesting them to confirm against a signature that they have read and understood the instructions. In addition, instructions are compiled in several languages

Vibrant mangement System 2015/1/13 44

6.Competence and training

TÜV NORD

Study Assessment

- How are staff familiarized with process safety issues? Are service providers taken into account? How is competence maintained?
-
- Acquiring an understanding of safety issues requires that new employees and service providers are well-trained. Refresher courses and further training should be arranged for existing employees as well.
- One way of taking care of staff training is to maintain a training plan and to monitor its progress. It is good practice to include service providers in the plan as well. It is also important to support competence-based training and training related to the development of work tasks.

Examples of good practice:

- The length of the orientation period is adjusted according to the task.
- Long absences are acknowledged in orientation practices.
- Staff training is carried out before extensive modification work.
- A training register is maintained.
- Training is a company-wide process.
- Written exams are administered to ensure that the training has been effective.
- Information from the training sessions is shared throughout the organization.

Vibrant management system 2015/1/13 45

7.Managing emergencies and deviations

TÜV NORD

Study Assessment

- What procedures does the company have in place in the event of emergencies and deviations?
-
- It is essential that staff know how to act during emergencies. For this reason, regular training in emergency procedures should be arranged. It is also recommended to include service providers in the training sessions.
- Staff and service providers should be encouraged to report near-miss situations. These reports help when it comes to identifying risks and determining what action should be taken where necessary. Potential accident scenarios should be investigated and the resulting data used to improve operations

Examples of good practice:

- A fire-drill group that trains weekly is formed.
- Department-specific fire drills are organized.
- An event is arranged where feedback on the success of an emergency drill is given. The drill is documented.
- A team of fire-fighters is maintained at the plant.
- Safety devices are tested and inspected regularly
- (e.g. safety lighting, safety showers, emergency eye-wash fountains).
- A 'chemical map' of the plant is created, including the location of chemicals, exits and fire extinguishing equipment.
- Container-specific extinguishing instructions are compiled.
- Close cooperation with the rescue services is maintained.
- Fire drills are arranged for all rescue service shifts.
- Rescue services from nearby regions are also included in the drills.

Vibrant Management sytem 2015/1/13 46

Include Process Safety in The Excellence Model

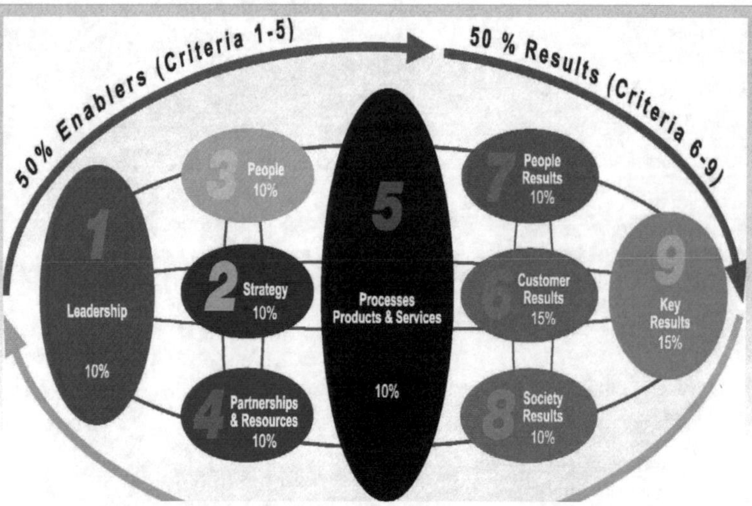

Leadership demonstrates Vibrant Management system to include Process Safety

Processes include Process Safety to earn Higher Grades

Conclusion

I believe that if measure of process safety performance is correctly implemented, it can lead us to be sure that there will not be another piper alpha or Macondo-type incident.

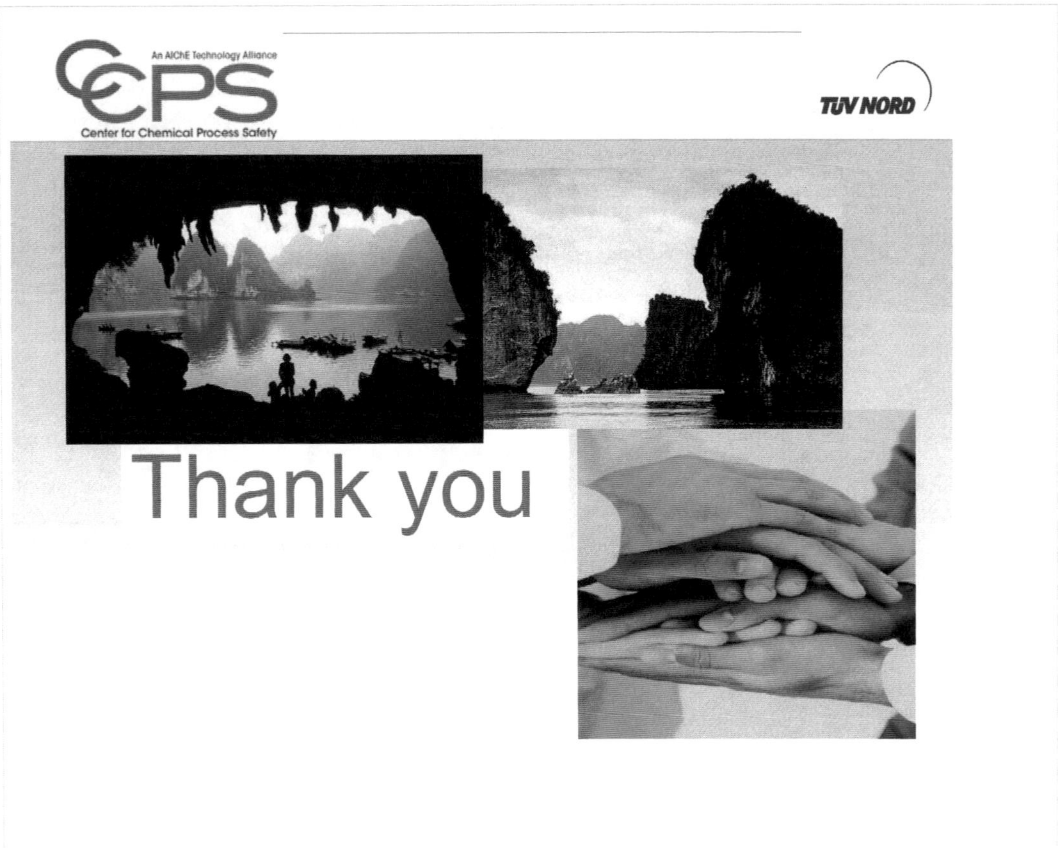
An AIChE Technology Alliance
CCPS
Center for Chemical Process Safety
TÜV NORD
Thank you

Embedding Safety in Design
Building a Sustainable Culture

Being Proactive- Not Reactive

PSM defined as....

"...Process safety refers to the prevention and mitigation of unintentional releases of potentially dangerous materials or energy...Process safety management involves a particular type of risk management - identifying and controlling the hazards arising from ...processes, such as the prevention of leaks, spills, equipment malfunctions, over-pressures, excessive temperatures, corrosion, metal fatigue, and other similar conditions."

Extract from the Report of the BP US Refineries Independent Safety Review Panel ("Baker Panel")

2

What Is Safety Culture?

- Culture is the tendency in all of us – and our organization - to want to do the **right thing** in the **right way** at the **right time, ALL the time** – even when if no one is looking
- Culture is the result of all the actions - **and inactions** - in institutional/workforce memory
- Culture is hard to measure and more difficult to change; it will be the "root cause of the decade"

3

Report of the BP US Refineries Independent Safety Review Panel ('Baker Panel')

Texas Refinery Incident 23 March 2005

Context

- 1997 getting Health, Safety and Environment right (gHSEr) established as our HSE framework
- 1999-2005 declining injury rates - among the best in the industry
- 23 March 2005 – Texas City incident
- BP and CSB investigated the incident
- Baker Panel established at recommendation of CSB in October 2005

5

Emphasis on Process Safety

- Process Safety Management requires an approach that is specific and distinct from personal safety or other short-term performance parameters in that:
  - Predictive indicators are essential to identify Process Safety issues because these take longer to observe
  - This requires a different approach that does not lend itself well to short term performance focus
  - It requires a disciplined, sustained approach and follow-up is essential
  - Everyone's activities can impact Process Safety

6

The Investigation Findings

- Process Safety Culture
- Process Safety Management Systems
- Performance Evaluation, Corrective Action
- Corporate Oversight

7

Process Safety Culture - Findings

- BP has not adequately established Process Safety as a core value
- BP has not provided effective leadership on or established appropriate operational expectations regarding Process Safety performance
- BP has emphasized personal safety but not Process Safety - mistakenly used improving personal safety performance (i.e., personal injury rates) as an indication of acceptable process safety performance ; BP's reliance on this data and inadequate process safety understanding created a false sense of confidence that it was properly addressing process safety risks at those refineries
- BP has not established a positive, trusting, and open environment with effective lines of communication between management and the workforce, including employee representatives
- BP does not have a designated, high-ranking leader for process safety dedicated to its refining business nor ensured that the resources required were identified and provided
- BP's corporate initiatives have overloaded personnel to the possible detriment of process safety

8

Evolution of PSM Standards

Process Safety has received a great deal of attention

- OSHA (Occupational Safety and Health Administration, US) promulgated a standard for Process Safety Management of Highly Hazardous Chemicals in1992 that has been used as an template for Process Safety Management around the world
- CCPS (Center For Chemical & Process Safety
- API-RP-750 (American Petroleum Institute)
- EPA 40 CFR (Environmental Protection Act
- OSHA 29 CPR 1910-119 (Occupational Safety & Health Administration
- RBPS – Risk Based Process Safety
- OISD – Oil India Safety Directorate
- Industries have adopted various models based on the these or a combination of PSM approaches

Process Safety Management – OSHA's 14 Elements

1. Work Permit System
2. Pre-startup Safety Review
3. Management of Change
4. Incident Reporting & Investigation
5. Emergency Planning and Response
6. Employee Participation in Safety Culture
7. Process Safety Information
8. Operating Procedures
9. Contractor Safety Management
10. Mechanical Integrity & Asset Reliability
11. Process Hazards analysis
12. Compliance Audits
13. Training
14. Trade secrets

Process Safety Management - Elements

Process Safety:

- Process Safety information
- Process Hazard analysis
- Management of Change
- Asset Integrity
- Operating Procedures
- Pre Start-up Safety Review

Common to Process & Personal Safety:

- Training and Performance
- Safe Working Practices – Permit To Work
- Contractor Safety Management
- Employee Participation in Safety Culture
- Audits and Corrective actions
- Incident Investigation
- Emergency Response
- Compliance Audits

Refining the PSM Model

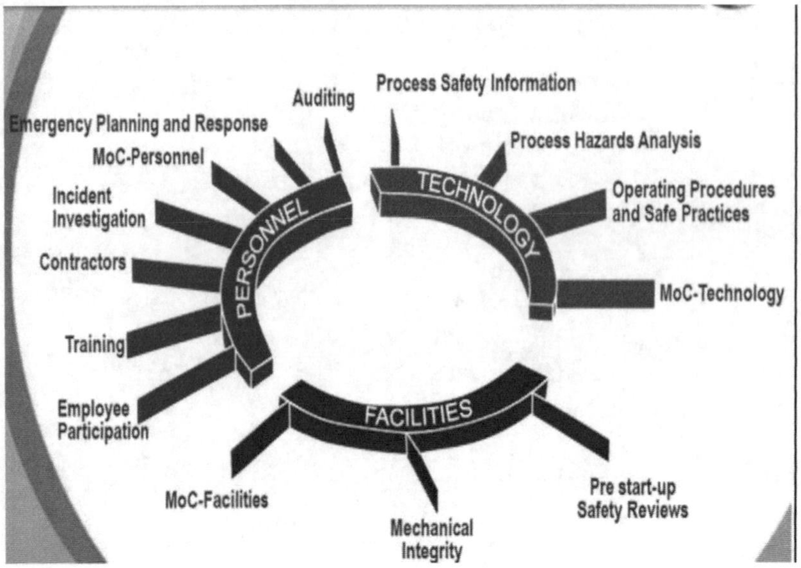

13

Process Safety Management - Approach

- Dedicated Cell with cross function specialists from Chemical, Mechanical & Electrical Background
- Co-opted members from other Teams.
- Led by TS & facilitated by Head-HSE (Energy Business).
- Specialized training by Learning Center.
- Detailed Gap Analysis on PSM Status at Refinery
- Develop time bound action Work plan for PSM implementation.
- Action Plan in coordination with Line Management involvement.
- Utilize expertise & experience of existing PSM Leads.
- Implementation as per action plan.
- Dissemination of PSM at plant level.
- PSM Audits on various systems.

PSM - Roadmap

- ✓ Focus on PSM implementation.
- ✓ Complete implementation pre commissioning
- ✓ Elementary – Phased rollout over 3 years
- ✓ Dedicated Team with full time involvement.
- ✓ Use of experienced & specialists functions.
- ✓ Integration Systems with OHSAS-18001.
- ✓ Involvement across plant level.
- ✓ QRA of existing facilities and new projects.
- ✓ FMEA of critical systems.
- ✓ Strengthened audit of systems.
- ✓ Integration of PSM with new / expansion projects at design stage.

Ownership of PSM Elements

	PSM Elements	Driven by
1	PSSR (Pre Start Up Safety Review)	Projects
2	SOP/SWP (Safe Operating Procedures)	Operations
3	ERP (Emergency Response Planning)	Fire
4	EPiSC (Employee Participation in Safety)	HR
5	PSI (Process Safety Information)	Tech Services
6	MoC (Management of Change)	Quality
7	PtW (Permit To Work)	Safety
8	PHA (Process Hazard Analysis)	Tech Services
9	MIAR (Mechanical Integrity)	Engg
10	CSM (Contractor Safety Management)	Projects
11	II (Incident Investigation)	Operations
12	T&C (Training & Communication)	Learning & Development
13	Compliance Audits	Safety

16

PSM Elements – Integrated in Year Business Plan

Design of the Employee Participation in Safety Culture by June & implemented fully by July 2008.
Business Owner – Head HR | Process Owner – R Shankar
Business Lead – HOD HR | Process Lead - Sushil Pandey

Design of the Process Safety Information system by June 2008 end & implemented fully by March 2009
Business Owner– Head Tech Services | Process Owner – R Shankar
Business Lead – HOD TS i | Process Lead – L R Dalsania

Review current Operating procedures against PSM standard and close gaps by Sept. 2008
Business Owner – Head Refinery | Process Owner – R Shankar
Business Lead – HOD Operations | Process Lead – Ritesh Shah

Design of the Contractor Safety Management system by May & implemented fully by June 2008.
Business Owner – Head Projects | Process Owner – R Shankar
Business Leads – HOD Projects | Process Lead - Siva Kumar/Suresh H

Design of the Mechanical integrity & Asset Reliability by July & implemented fully by Oct 2008.
Business Owner – Head Maintenance | Process Owner – R Shankar
Business Lead – HOD Reliability | Process Lead – Prashant Shah

All PSM Elements Business Leads – Area Representatives | Process Leads – Area PSCs

Status on Key PSM Improvement through KPIs

KPI	Target	Status	Remarks
Design & implement Process Safety Information system	Mar 09	70%	
Design & implement Mechanical Integrity & Asset Reliability	Oct 08	66%	
Design & implement Contractors Safety Management	June 08	100%	
Review SOP against PSM & close gaps	Sept 08	100%	
Design & implement Employees Participation in Safety Culture.	July 08	100%	

PSM Elements Progress Monitoring

PSM Element	Status
Permit to Work System	Work Permit Procedures implemented and PTW System in place
Pre Start-up Safety Review	PSSR fully implemented and Format system in place
Management of Change	MoC online on a web platform designed .
Incident Reporting & Investigation	Incident Reporting & Investigation System implemented on a fully - Web platform, RCA Tool and action tracking
Emergency Planning & Response	On – Site Emergency Plan in place with regular tests

Benefits of PSM

- ✓ Preventing Hydrocarbon fires, explosions and toxic gas releases.
- ✓ Improved HSE review of new projects.
- ✓ Assessment of Risks by QRA.
- ✓ Effective controls & emergency response planning.
- ✓ Positive effect on Employee morale at work place and the community
- ✓ Absence/inadequate procedures escalation ahead of time
- ✓ Compliance with industry codes, standards and practices.
- ✓ Integrates with existing quality, environmental and safety management systems.
- ✓ Reduces downtime and losses.
- ✓ Viewed favorably by external interested parties like Factory Inspectorate and Pollution Control Boards.

PSM @ Healthcare

Pharma Industry

- **API**
- **Formulation**

Operational Hazards

API

- **Handling Hazardous Chemicals**
- **Carrying out Hazardous Reactions**
- **Using large quantity of Solvents**
- **Handling explosible Powders**

Formulations

- **Handling of explosible Powders**
- **Operating tableting/filling and packaging machines**

21

PARIVARTAN

A Cultural Transformation Journey

An Overview

Starting with a Safety Perception Survey

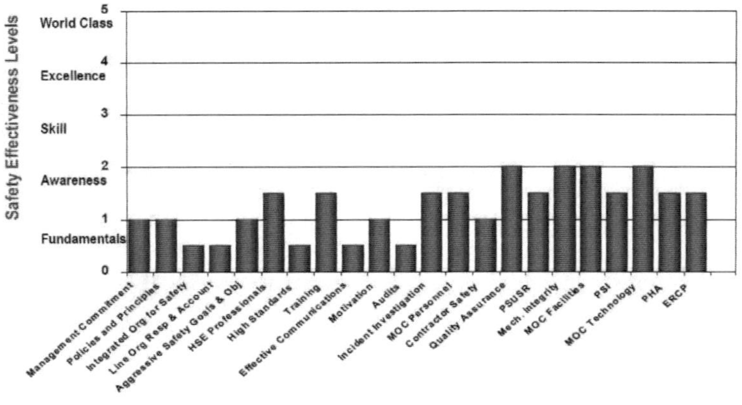

Parivartan Roadmap

Support Framework - 5 Step Process

The support for 'safety cultural transformation' would follow a **five** step process:

1. **Quick Wins through Containment Phase**
2. **Engaging Leadership & Line Organization**
3. **Consistent Implementation through Work Streams**
4. **Enhancing Safety Management System & Competency**
5. **Operating Discipline for Sustained Performance**

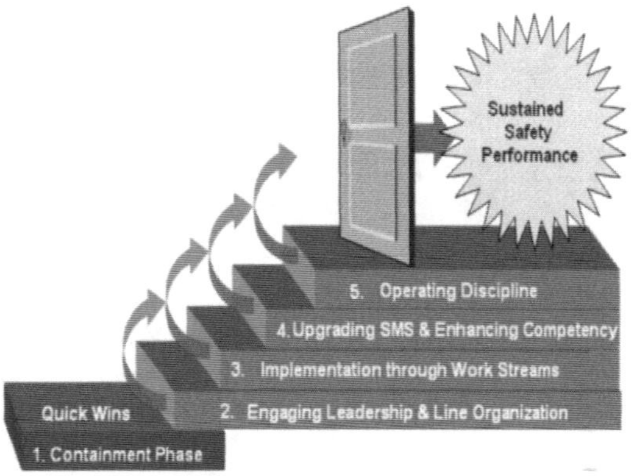

Fig: Pictorial representation of five step process

DuPont Safety Management System

Consists of 22 Elements

- ➢ Behavioural Safety Management 12 Elements
- ➢ Process Safety & Risk Management 14 Elements

(4 elements of Behavioural Safety overlapped in PSRM)

25

DuPont SMS - 22 Elements of Safety

- The DuPont model contains 22 elements required to achieve both safety leadership and operational excellence
- Management commitment at the core
- Includes cultural and risk-based elements

26

PSM : the DuPont Wheel

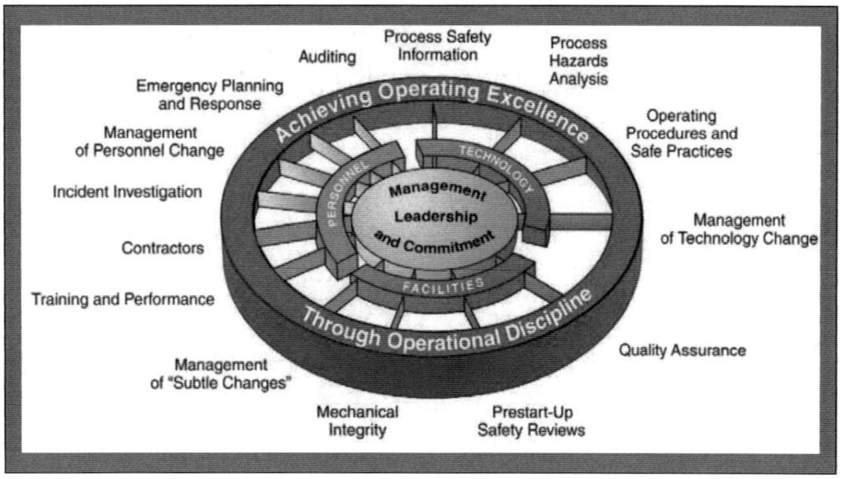

Parivartan through Work Streams

Engagement Model based on Six Work streams and Four Dimensions of change

Line Management Ownership

Proposed Integrated Safety Organization.....

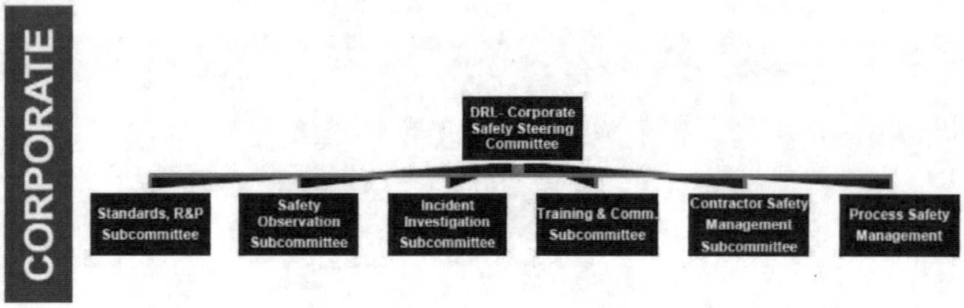

This will be mirrored at BU / Site level

Cultural Transformation - Essential Elements

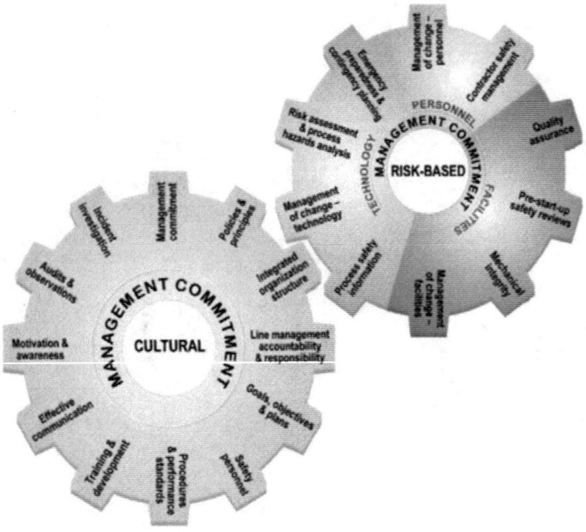

Management Commitment
Policies and Principles
Integrated Organization Structure
Line Management Responsibility
Goals, Objectives and Plans
Safety Personnel
Procedures and Performance Standards
Training and Development
Effective Communication
Motivation and Awareness
Audits and Observations
Incident Investigation
Contractor Safety Management

Creating an open and collaborative culture

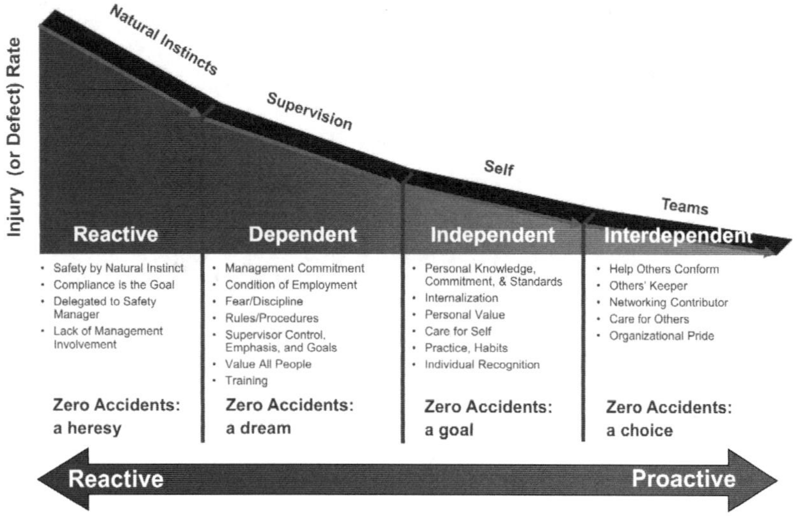

Challenges in Implementing Process Safety

- Ineffective EHS/PSM system performance
- Inadequate reaction to fixing identified problems - lack of follow-up/huge backlogs
- Superficial audits - "check the box" mentality
- Inadequate metrics-misplaced confidence in injury rates
- Poor management review at all levels

32

Challenges in Implementing Process Safety

- Weak conduct of operations/lack of operating excellence
- Poor incident reporting, learning and risk review
- No MOC (people, policies, or organization)
- Chronic cost-cutting and production pressure/ emphasis over process safety
- Plenty of "talk", but few examples of leading by doing

33

Process Safety Culture – Essential Features

Establish safety as a core value	**Defer to expertise**
Provide strong leadership	**Ensure open and effective communications**
Establish and enforce high standards of performance	**Establish a questioning/learning environment**
Formalize the safety culture emphasis/approach	**Foster mutual trust**
Maintain a sense of vulnerability	**Provide timely response to safety issues and concerns**
Empower individuals to successfully fulfill their safety responsibilities	**Provide continuous monitoring of performance**

34

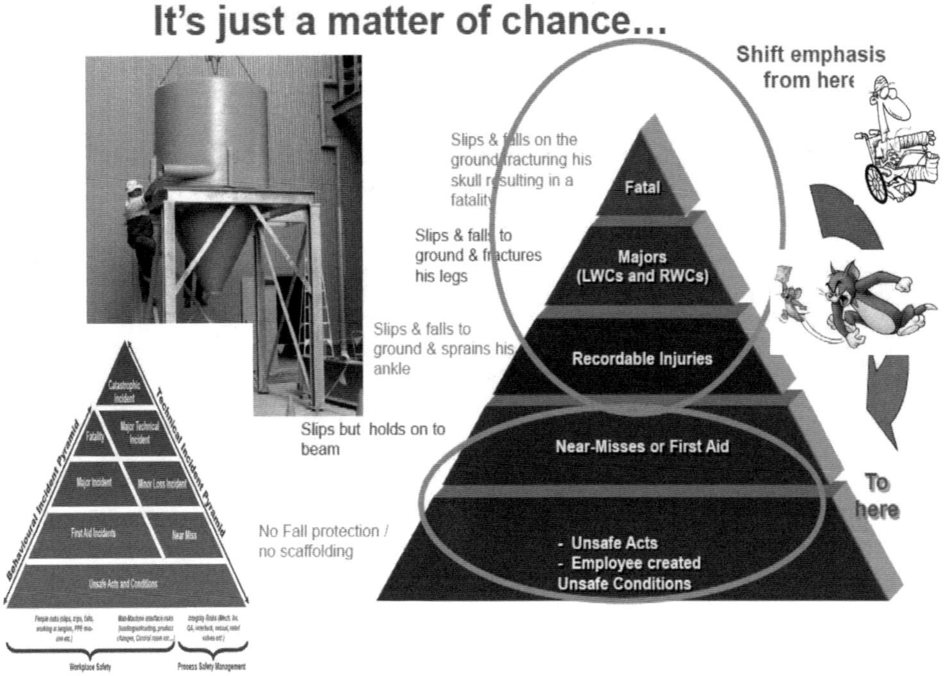
It's just a matter of chance...
Shift emphasis from here
Slips & falls on the ground fracturing his skull resulting in a fatality
Fatal
Slips & falls to ground & fractures his legs
Majors (LWCs and RWCs)
Slips & falls to ground & sprains his ankle
Recordable Injuries
Slips but holds on to beam
Near-Misses or First Aid
To here
No Fall protection / no scaffolding
- Unsafe Acts
- Employee created Unsafe Conditions
Behavioural Incident Pyramid
Technical Incident Pyramid
Catastrophic Incident
Fatality
Major Technical Incident
Major Incident
Minor Loss Incident
First Aid Incidents
Near Miss
Unsafe Acts and Conditions
Workplace Safety
Process Safety Management

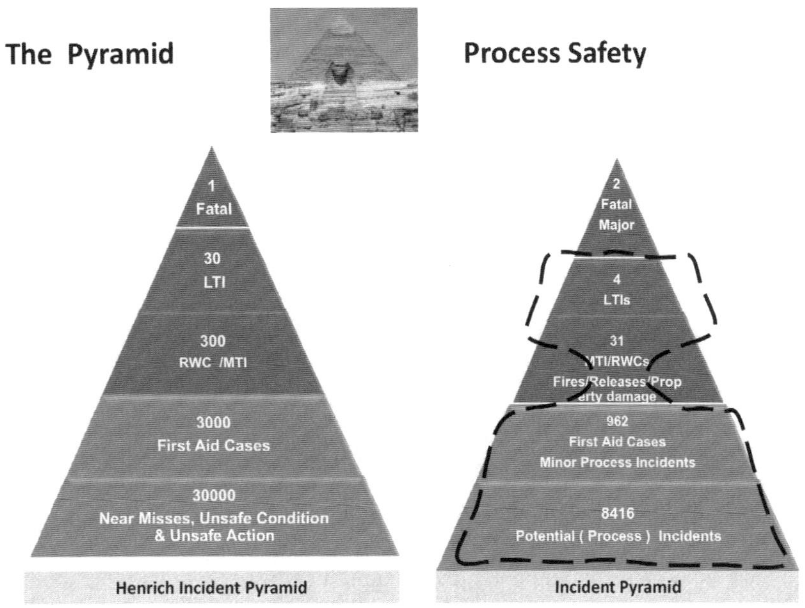
The Pyramid
Process Safety
1
Fatal
30
LTI
300
RWC /MTI
3000
First Aid Cases
30000
Near Misses, Unsafe Condition & Unsafe Action
Henrich Incident Pyramid
2
Fatal
Major
4
LTIs
31
MTI/RWCs
Fires/Releases/Property damage
962
First Aid Cases
Minor Process Incidents
8416
Potential (Process) Incidents
Incident Pyramid

3 Questions to ask before any activity

What could go wrong with what I am about to do?
मेरे काम के साथ क्या गलत हो सकता है ?

😐

How could this hurt me?
काम करते वक्त मुझे चोट कैसे लग सकती है?

☹

What must I do to avoid getting hurt if it goes wrong?
काम करते वक्त मुझे चोट से बचने के लिए क्या करना चाहिए?

☺

If anything can go wrong, it will ! If anything can be done right, you CAN DO it right !!

Questions

Thanks for your time !

shankarrajag@gmail.com

UNEP Approaches to Safer Production

Ms. Johanna Suikkanen, Programme Officer,
Business and Industry Unit, UNEP-DTIE

Presentation outline

1. UNEP's Safer Production Programme
   - APELL
   - Responsible Production
   - Flexible Framework Initiative for CAPP
2. Moving forward

UN Environment Programme

- Delivers support in 7 thematic priorities in **partnership** with IGOs, public sector, private sector, and civil society

- Functions include assessing the state of the world's environment, promoting global consensus and stimulate solutions to environmental problems, and build capacity and networks to facilitate implementation

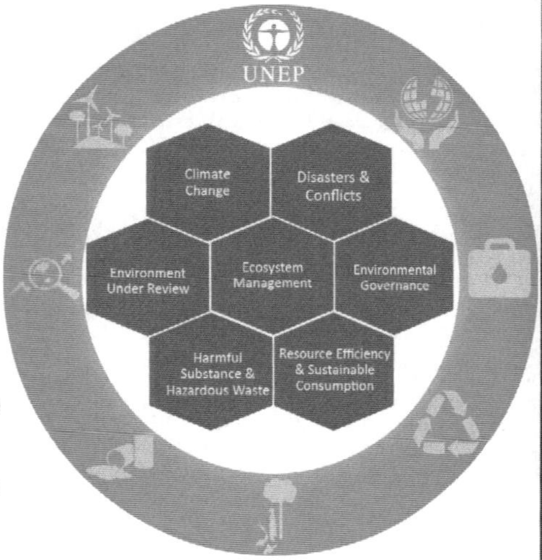

UNEP

UNEP Partnerships

- Multi-stakeholder partnerships:
  - With UN-system partners
  - With international and regional organisations
  - National governments and institutions
  - Universities
  - Industry associations and companies
  - RECP-net

- Allows to leverage for impacts:
  - Up-scaling UNEP initiated pilot activities
  - Building capacity
  - Technical expertise and support
  - Sharing expertise and synergizing effort

UNEP

UNEP and Industry

- Promotes a **sustainable industrial development** that decouples economic growth from the consumption of natural resources

- Partners with governments and industry in promoting sustainable consumption and production:
  - Cleaner Production and resource efficiency
  - Safer Production
  - Corporate Social Responsibility and Corporate Reporting
  - Eco-innovation

5

2014 CCPS Global Summit on Process Safety
Responsible Collaboration Session

UNEP Safer Production

UNEP provides support to governments and the private sector on awareness, prevention, preparedness and management of industrial risks

- Following a series of accidents including Seveso, Bhopal, Mexico City and others

- Requested by UNEP Governing Council in 1986 to work in partnership with governments, industry and other UN bodies to establish a programme that enables multi-stakeholder cooperation to minimise risks

- Led to the establishment of the **APELL** programme

6

2014 CCPS Global Summit on Process Safety
Responsible Collaboration Session

UNEP Safer Production Portfolio

National Level: Flexible Framework Initiative for CAPP

- Guidance for national policy development for chemical accident prevention and preparedness http://capp.eecentre.org/

Local Level: APELL

- Local level preparedness for emergencies http://apell.eecentre.org/

SMEs: Responsible Production

- Risk management and communication along the value chain - aimed primarily at SMEs http://www.unep.org/responsibleproduction/

UNEP

UNEP Partnering for Safety

- Improving safety requires coordination, collaboration and consultation. UNEP facilitates this by:
  - Building partnerships and open dialogue
  - Establishing expert networks to share best practice and develop guidance
  - Promoting partnerships at country level to faciliate stakeholders to work together
  - Establishing partnerships with organisations that bring in different expertise to support capacity building globally

UNEP

APELL

- **Awareness & Preparedness for Emergencies at Local Level**
- A multi-stakeholder approach involving industry, local community leaders and local authorities
- **Objectives**:
  - Risk awareness, communication and community education
  - Coordination between representatives from the industry, the local level institutions and the public
  - Improved emergency preparedness planning
  - Measures for risk reduction, accident prevention and mitigation

10

APELL

- Developed in partnership with the US EPA, CEFIC and the international chemical industry
- APELL Handbook published in 1988
- Sector applications that have been developed to date:
  - TransAPELL
  - APELL for Port Areas
  - APELL for Mining
  - APELL for Coastal Tourism Destinations
  - APELL for multi-hazards

APELL

12

Responsible Production

- Company focus: specifically targets **SMEs**
- Addresses the need for simplified guidance on chemical hazard management for on-site and off-site operations
- '***Responsible Collaboration***' is at the heart of Responsible Production:
  - Integrates PSM, Corporate Social Responsibility and APELL in one framework
  - 5-step systematic methodology with a scope extending beyond the individual site
    - Step 2: Stakeholder identification and engagement

13

Responsible Production

- Developed in partnership with industry (ICCA, ICMM), and Public institutions (Peru, Thailand)
- **Responsible Production- A Framework for Chemical Hazard Management in SMEs (2010)**
- Responsible Production Learners and Trainers Companion
- Responsible Production Online Interactive tool
- **RP for Chemical Hazard Management - Lessons Learned from Implementation (2013)**
  - Tells the 'story' of RP through Pilot case studies
  - Feedback and recommendations for continuous improvement

UNEP

Responsible Production

Extends chemical safety management across the value chain:

- Understanding the risks and related concerns of the people who will be affected by the SMEs' products, actions and decisions, or whose own actions and decisions can affect the SME and its performance.
- Interesting not only for SMEs, but also large companies doing business with SMEs

UNEP

Responsible Production: 5 steps

UNEP

1. Identify responsible production issues
2. Get the right people involved
3. Develop your plan
4. Put the plan into practice, train and communicate
5. Evaluate how well you did

Typical challenges for SMEs:

- Labelling
- Housekeeping
- Use of appropriate PPE
- Use of SDS
- Seeing risks outside of core operations

Flexible Framework Initiative

UNEP

- Prevention and Preparedness
- Builds on 30 years of experience

International Conventions (e.g., ILO 174, UNECE TEIA)
International programmes (e.g.OECD's chemical accidents programme)
Legal instruments (US RMP, European Seveso directive)
Chemical management approaches (e.g. SAICM and OECD Guiding principles)

Prevention of accidents
Preparedness for accidents
Elements of CAPP programmes

saicm

Flexible Framework Initiative

- **A Flexible Framework for Addressing Chemical Accident Prevention and Preparedness: A Guidance for Governments**
  - How to develop a CAPP programme
  - Elements of a CAPP programme
  - Outlines Roles of Industry and Roles of Authorities in a CAPP Programme

- **A Flexible Framework for Addressing Chemical Accident Prevention and Preparedness: An Implementation Support Package**
  - Practical guidance for initating multi-stakeholder collaboration in addressing CAPP

Flexible Framework Initiative

- Emphasises the interrelated responsibilities related to CAPP:
  - Role of Authorities: Prepare, impose and enforce obligations
  - Role of Industries: General obligation to operate safely, assess and manage risks, risk communication

- Encourages inter-agency and multi-stakeholder collaboration in the process of developing policies and programmes (eventually regulations) for CAPP
  - An inter-agency multi-stakeholder task force drives the development of a national CAPP Programme

- Case: Good example of cross country collaboration occurred between Sri Lanka and Thailand

Online resources

- More information, case studies and tools on:
  - Flexible Framework Webportal http://www.capp.eecentre.org/
  - APELL http://www.apell.eecentre.org/
- Online Course on Industrial Accidents
  - A 3 hour basic course
  - Hosted by the Environmental Emergencies Centre: www.eecentre.org
- APELL Youtube Channel
- RP interactive tool

Lessons learned

- UNEP experience: APELL in 30 countries (80 + communities), Responsible Production in 10 + countries, and Flexible Framework for CAPP in 6+ countries:
  - **National level**: Leadership at high level, with multi-stakeholder processes lead to outcomes
  - **Local level:** Locally relevant 'vision of success' and locally relevant activities (e.g. risk communication, training)
  - **SMEs:** Lack of technical capacities on chemical safety, and experience in working with stakeholders; importance of supply chain influence and external technical support

Thank you for your attention!

Contact details:

Ms. Johanna Suikkanen
Programme Officer (Cleaner and Safer Production)
UNEP-DTIE
johanna.suikkanen@unep.org
+33 (0)1 44 37 19 84

GREEN TRIANGLE FOR SAFETY ®

NATIONAL SAFETY COUNCIL INDIA

1

GREEN TRIANGLE FOR SAFETY ®

NSCI EXPERIENCE IN APELL IMPLEMENTATION

V. B. Sant
Director General
National Safety Council, Navi Mumbai

ABOUT NSC INDIA .....

- Established by Government of India on 4th March 1966
- Autonomous, independent, non-commercial, non-profit making society
- Membership base of more than 8750 comprising more than 5570 companies
- 51 members multi-partite Board of Governors
- Eminent industrialist as Chairman, two vice-chairmen
- 18 chapters covering 85% geographical area

NSCI SERVICES

EVOLUTION OF APELL PROCESS IN INDIA

1992-97*	APELL-LAMP Project in six high risk industrial areas
2000-02	Trans-APELL (demonstration project) in Trombay-Chembur area
2002-04	Establishment of National APELL Centre
2004 and 2006	Launching of NAC sub-centres at Tuticorin TN and Pune MAH
2006-08*	Disaster Risk Reduction in Tourism Destination (Kanniyakumari)

APELL-LAMP PROJECT(1992-97) (UNEP/DTIE, USAID, WEC)

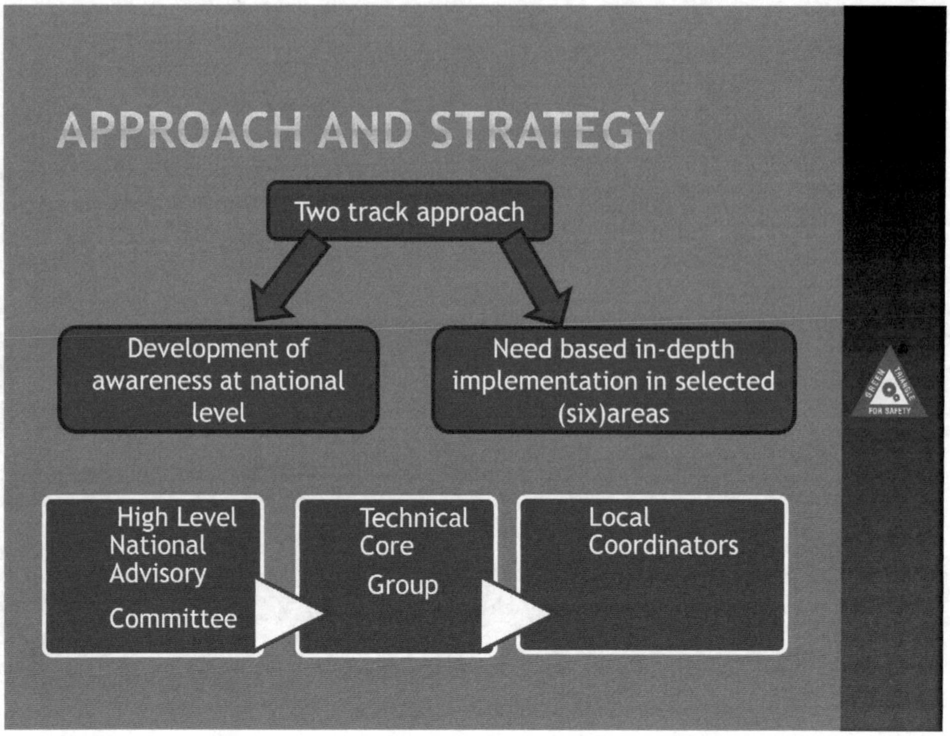
APPROACH AND STRATEGY
Two track approach
Development of awareness at national level
Need based in-depth implementation in selected (six)areas
High Level National Advisory Committee
Technical Core Group
Local Coordinators

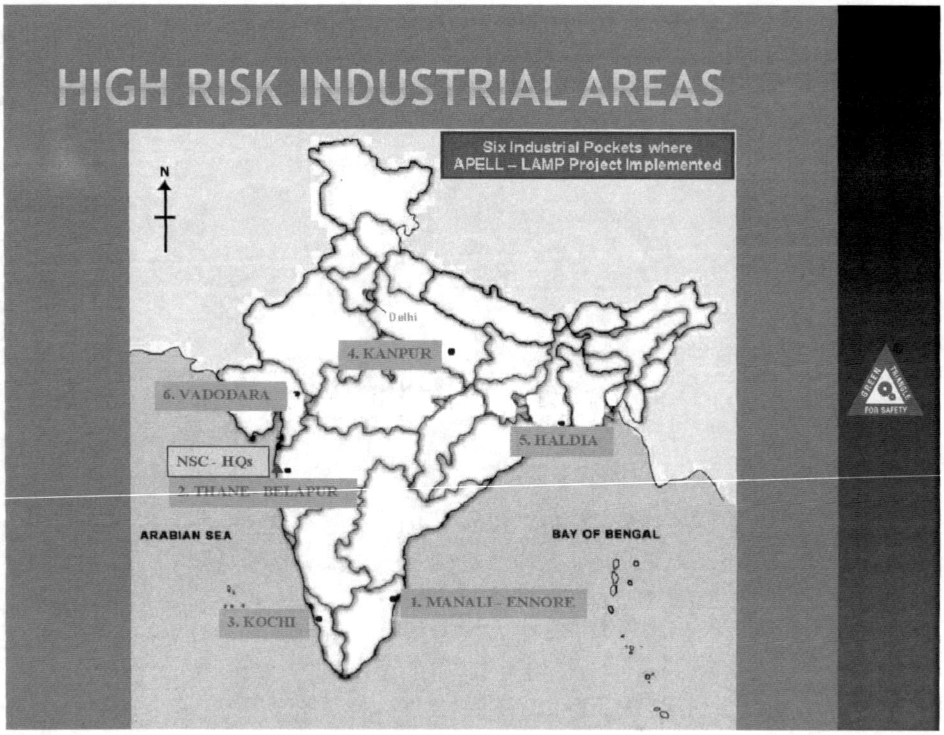
HIGH RISK INDUSTRIAL AREAS
Six Industrial Pockets where APELL – LAMP Project Implemented
N
Delhi
4. KANPUR
6. VADODARA
5. HALDIA
NSC - HQs
2. THANE - BELAPUR
ARABIAN SEA
BAY OF BENGAL
1. MANALI - ENNORE
3. KOCHI

OVERVIEW : ACTIVITIES

Workshops	16	994
Seminars	5	181
Training Courses	6	178
Hands-on-training	5	136
Exercise/drills	4	-
Audits	2	-

MAJOR OUTCOMES

- **Notification of "Chemical Accidents (Emergency Planning, Preparedness and Response) Rules, 1996"**
- **Crisis Groups at the national, state, district and local levels on the model of APELL Coordinating Group set up**
- **APELL recommended in the National Guidelines on Chemical (Industrial) Disaster issued by NDMA**

DISASTER RISK REDUCTION PROJECT- KANNIYAKUMARI

APPROACH AND STRATEGY

- Constitution of CG
- All stakeholders as members of CG
- Role of NSC
- Appointment of Local Coordinator

ACTIVITIES

- Workshops for training of CG members
- Studies
- Training of local authorities
- Preparation of documents
- Information booklets/leaflets
- Seminars

OUTCOMES

- Development of Disaster Management Plan for Kanniyakumari Town
- Training of responders
- Emergency drill

EMERGENCY DRILL

EMERGENCY DRILL

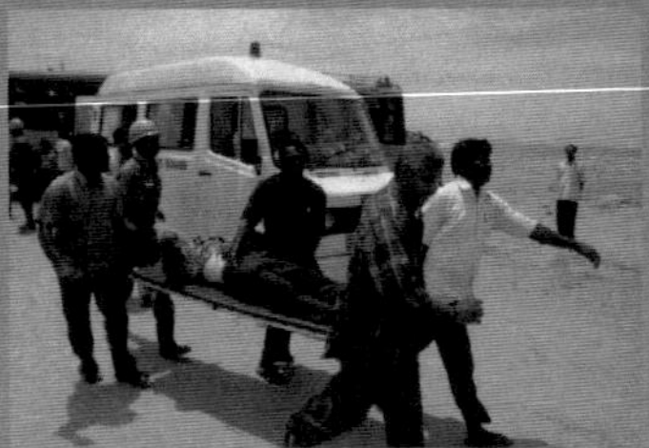

LIMITATIONS IN IMPLEMENTATION

- Overlapping constitution and role of Local Crisis Group (LCG) and Coordinating Group (CG)
- LCG not effective
  - Lack of clarity - roles & responsibility of LCG members
  - Absence of exclusive and dedicated manpower
  - Lack of funds for activities of LCG
- Absence of national level technical organization for coordinating, guiding and monitoring the activities of LCGs

CASE STUDY
FIREWORKS FACTORY ACCIDENT IN SIVAKSI TAMIL NADU INDIA

THE ACCIDENT

- First explosion at 11.30 am on 5th September 2012
- Followed by many low intensity explosions
- Massive explosion at 1.15 pm

ACCIDENT SEQUENCE

- No of workers inside - 210
- Many of them working in open spaces
- One of the workers applied excessive pressure while ramming the explosive
- Instantaneous explosion, worker thrown away and others injured
- All workers ran away taking injured with them
- Many explosions and fire
- Villagers and families gathered around-about 300
- Massive explosion at 1.15 pm

CONTRIBUTING FACTORS

- Working in open spaces
- Use of prohibited implements
- Engaging unskilled and un-trained workers
- Exceeding the stock of explosives
- Exceeding the manpower limit
- Gathering of large number of onlookers in and around factory premises

FINDINGS OF INQUIRY

- 40 persons dead including 35 onlookers
- 89 injured
- Lack of coordination amongst the authorities
- Lack of public awareness major cause of injuries to many onlookers

Involvement of community and all stakeholders

LESSIONS LEARNT-APELL

- Full time emergency coordinator
- Initiative by Local industries
- Shelter places
- Public communication channels
- First responders should be trained and equipped adequately

THANK YOU ....

Approaches for Improved Stakeholder Collaboration on Chemical Accidents in Sri Lanka

Samantha Kumarasena

Chief Operating Officer

National Cleaner Production Centre,

Sri Lanka

Contents

- **Why** National Cleaner Production Centre, Sri Lanka did embark on Chemical safety activities
- Approaches used in **enhancement** of chemical safety at enterprise level
- **Lesions learned** from Responsible Production activities in Sri Lanka – in a nutshell
- **Vital aspects** of Stakeholder mapping, involvement and risks aversion
- Approaches for **Scaling up** chemical safety in SMEs

1/13/2015

Economy & Industry Rationale

- Sri Lankan economy has reported the **highest growth** of GDP in South Asia in 2013 (6.3%)
- **Chemical use industries** have become predominant in the composition of Sri Lankan industry
- **Lack of know-how** in chemical use industries on how to deal with chemicals
- Chemical and process safety issues were **ignored** until the recent past

1/13/2015 3

Chemicals related industrial accident and pollution incidents were rising up

Eye-opening incident was in the latex dipping industry – community rising up against water pollution!
3 deaths over the military attack to protesters (2013)

1/13/2015 4

Decision of NCPC, Sri Lanka to embark on chemical safety approaches

- **Poor chemical safety performance witnessed** during field assessments of RECP activities
- **Increased chemical safety related incidents reported,** with the growth of economy /industry in the post-war period
- **Demand from the existing clients** for technical support and advisory services for improving safety aspects of the organization
- Vacuum created by **non-existence of national level service provider** for advising on chemical safety

1/13/2015 5

Approaches used by NCPC Sri Lanka for promoting Chemical safety

- implementation of **Chemical Leasing** in Sri Lanka since 2008 (Support by UNIDO)
- Pilot application of **Responsible Production** since 2010 (Support by UNEP)
- Hosting the local chapter of Responsible Care – **Lanka Responsible Care Council** since 2010 (Support by Newzealand RC council & ICCA)
- Supporting for **Chemical Accident Prevention Programme** of CEA since 2013 (Support by UNEP)

1/13/2015 6

UNEP Regional Training on RERP 2010

Application of Chemical Leasing (ChL) in Sri Lanka

- More than 15 pilot application projects since 2008
- Won Global Chemical leasing Awards held in bi-annually in 2012 and 2014

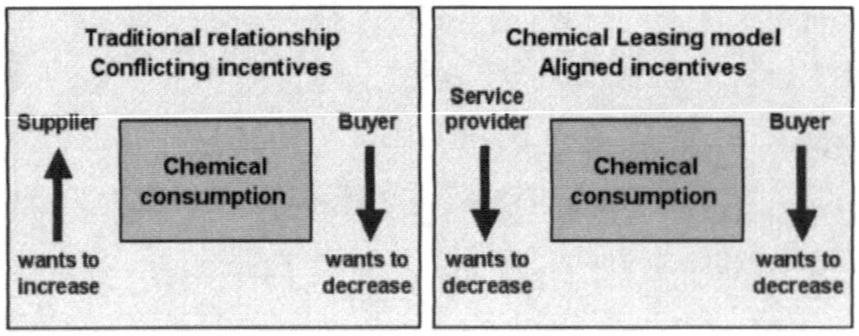

Case Study from Agriculture sector

- First successful application in the world
- Results of application in Cultivation of Carrot in Nuwara-Eliya, Sri Lanka

	Control plot	Chemical leasing
Total cost LKR	321,960	309,532
Revenue LKR	677,645	700540
Income over expenditure	355,785	391,007

Lanka Responsible Care Council

- Formed in the 2010 and received the membership in Global Responsible Care initiative in 2012
- Representing Major chemical suppliers and users (Multinational & Local)
- Focus Areas:

Responsible Care ,Workplaces Safety, Advocacy, Industrial and Agrochemicals Emergency preparedness Management

Common safety improvements under Cases of Responsible Production

- Responsible Production helped SMEs to understand and implement safety measures at firm level

Before the RP assessment

After the RP assessment

Proper use of MSDS

- Collection of all MSDS, displaying essential instructions in local languages at points of use
- prepare **operating procedure guidelines** and permits to work (PTW)

Reviewing and fixing PPEs

- **Conducting PPE risk assessment and providing appropriate PPE** as per the corresponding MSDS **and enforcing use of them**
- **Providing comprehensive PPE training** for employees

1/13/2015 13

Improved housekeeping

1/13/2015 14

Introducing best practices

- **Hazop study** to be conducted before purchasing equipment
- **Emergency retrieval mechanism** was introduced to withdraw the workers especially from reactors and other confined spaces/vessels
- **Avoid Drilling work etc. very close to reactors** without following 'permit-to-work' safety procedures

1/13/2015 15

Minimizing risk of nitro-cellulose storage

Old Nitro-cellulose storage

New storage with all safety

1/13/2015 16

Improving the safety infrastructure

Safety measures were implemented for

- Avoiding the accumulation of static charges and avoiding sparks
- Periodic checking of safety systems in place
- Fixing guards for movable components, hot surfaces etc.
- Maintaining the built containments
- Infrastructure for fire water handling

Transportation

- **Design transport emergency (TREM) cards,** properly filled and endorsed, and provide with the transporters
- **Label** bags, containers, etc. containing chemicals **clearly and visibly**

Warning boards are on display in the material storage after RP

Stakeholder Mapping: E.g. Stakeholders of the Rubber industry

	Relevance			Importance
	Yes	No		
Suppliers (Local)				
Supply of CaCO3 & Clay	X		Lanka Minerals & Chemicals (Pvt) Ltd	3
Supply of CaCO3	X		Inmine (Pvt) Ltd	3
Local Agent for supplying of antioxidants, pigments, accelerators, silicates, and activators, blowing agents, vulcanizing agents, etc.	X		(i) Multiway (Pvt) Ltd, Colombo (ii) Chemicals & Equipment (Pvt) Ltd	3
Supply of furnace oil	X		Ceylon Petroleum Corporation	1
Supply of processing oils	X		CSI International (Pvt) Ltd	1

Stakeholder Mapping: E.g. Stakeholders of the Rubber industry- Contd...

	Relevance			Importance
	Yes	No		
Buyers (Local)	X		D Samson & Sons (Pvt) Ltd	4
Local authorities	X		Pradeshiya Sabha, Akmeemana	3
Government	X		Central Environmental Authority	3
	X		Labour Department	3
Utility Suppliers	X			1
Water	X		Water board (NWSDB)	2
Electricity	X		Ceylon Electricity Board	4
Community groups			Residents in the area	2
			Self employed families	1

1/13/2015 211

Stakeholder Mapping: E.g. Stakeholders of the Rubber industry - Contd...

	Relevance			Importance
	Yes	No		
Employees (employee representatives)	X		Staff (Management / Executive level)	4
Transporters	X		Operator level	3
)Transport of $CaCO_3$ & Clay	X		Lanka Minerals & Chemicals (Pvt) Ltd	2
(b) Transport of $CaCO_3$	X		Inmine (Pvt) Ltd, Balangoda	2
(c) Transport of imported chemicals	X		Galle Stores (Pvt) Ltd – Welisara	1
Retailers	X		ODEL	1
	X		No Limit	1
1/13/2015			Fashion bug	1

Stakeholder Mapping: E.g. Stakeholders of the Rubber industry - Contd…

	Relevance			Importance
	Yes	No		
Insurance Company	X		Sri Lanka Insurance	4
Social & Religious Groups in the area	X		Clergy (Temple)	3
Service Providers	X		Food suppliers	1
	X		Waste collectors	1
	X		Sub contractors	1
	X			

1/13/2015

Stakeholder mapping in companies of the value chain of paint industry sector

Stakeholder Group	C1		C2		C3		C4		C5		C6		C7	
	R	I	R	I	R	I	R	I	R	I	R	I	R	I
Community	N	-	Y	4	Y	2	N	-	Y	4	Y	4	Y	2
Neibouring industries	Y	2	Y	2	Y	2	Y	4	N	-	Y	3	Y	4
Buyers/Customers	Y	4	Y	4	Y	4	Y	4	Y	4	Y	1	Y	4
Suppliers	Y	2	Y	1	Y	1	Y	2	Y	2	Y	4	Y	2
Transporters	Y	2	Y	2	Y	1	Y	2	Y	1	Y	2	Y	2
Service providers	Y	1	Y	1	Y	1	Y	3	Y	1	Y	2	Y	2
Employees	Y	4	Y	4	Y	4	Y	4	Y	4	Y	4	Y	4
Sub contractors	Y	2	Y	1	Y	2	Y	2	Y	2	Y	2	Y	2
Government authority	Y	3	Y	3	Y	3	Y	4	Y	3	Y	4	Y	4
Social and religious groups	Y	2	Y	4	Y	1	Y	4	Y	4	Y	4	Y	4

24

Methods of engagement of stakeholders

Methods of engagement of stakeholders are different among stakeholder groups; in common

- ✓ Awareness programmes/briefing sessions / training
- ✓ Group discussions and feed back
- ✓ Advisory panels
- ✓ Work place assessments
- ✓ Drills

1/13/2015 25

Methods of engagement of stakeholders

- ✓ Monitoring and periodic evaluations
- ✓ Rewarding schemes for best suggestions and performances
- ✓ Leaflets, Bulletins, brochures, video clips
- ✓ Invited Speeches, presentations
- ✓ Study tours

etc.

1/13/2015 26

Risks of involvement of stakeholders

Stakeholder	Risk	Risk rating
Community	Rising up against the company operations	Very high
Neibouring industries	Rising up against the company operations	moderate
	Divulging proprietary company information	High
Buyers/Customers	Divulging undisclosed company information	High
Suppliers	Divulging undisclosed company information	High
Transporters	Unnecessary fear on company operations and miscommunication	Low
Service providers	Unnecessary panic situation created and miscommunication	Low
Employees	Unnecessary panic situation created	High
Sub contractors	Unnecessary fear on company operations	Moderate
Government authority	Unnecessary fear on company operations	Very low
Social and religious groups	Rising up against the company operations	High

1/13/2015 27

Stakeholder involvement while managing the risks

- **Implementing** the risk mitigation plan is the **baseline**
- Proper education of **employees** and using them as **workhorses**
- Developing good **public relations** with the community, social and religious groups
- Starting with welfare
- **Step by step** and **Group by group** involvement

1/13/2015 28

Society as a sources of learning

- How **Corporate** manage in other vital areas
- Spirits and values of the **Society**
- Public relations, personnel relationships and networking with **people**
- **Success stories** of converting public protests into pillars of social security

Stakeholder involvement with step by step approach

- Developing relationships by attending to the public events and individual's functions
- Inviting the religious and community leaders to factory for outdoor events
- Involving the community in outdoor functions
- Winning the 'hearts' – trust and confidence
- Educating the community on Risk mitigation measurers and the level of operational controls

Stakeholder involvement with step by step approach

- Taking them for plant visits and demonstrations
- Cross –plants internal audits within the Industrial estate
- Safety Audits in the plant by community leaders
- Involving habitants living in the effective radius of risk of hazard in the mock drills

1/13/2015 31

Key positive responses from the companies

- **Improvement in chemical safety was recognized as the need of the hour** during early stage of training itself
- **Knowledge gap and inadequate in-house capacity of the companies was realized** when training and initial work-through was conducted
- **Motivation among floor level employees** was very high for RP activities – **became a driver**

1/13/2015 32

Crucial negative responses from the companies

- **Concerns on confidentiality of information specially of paint companies** was a real barrier – not willing to accommodate external experts for field level assessments
- **Lack of preparedness for Risk communication** (to the community or neighborhood) was the main barrier to overcome

1/13/2015 33

Top down strategies to **promote** chemical safety at work place and factory surroundings

- **Policy level measures** are suggested to improve the public image of so called **high polluting industries**
- **Legal instruments** to be developed **to enforce the chemical safety**
- More **assistance to the industries** to improve chemical safety via preparedness and prevention
- Recognized RP as a tool for improving chemical safety at firm level in the **Road map for CAPP**

1/13/2015 34

Bottom up strategies

- **Package and link up chemical RP with Resource productivity and environmental management tools**
- Promoting RP **as a tool to comply with Responsible Care guiding principles** of Lanka Responsible Care Council (LRCC)
- Introduce as a **ground preparing tool for OHSAS 18001**
- Position as a **methodology for Chemical Accident Prevention in hazardous installations (CAPP)**

1/13/2015 35

Conclusion

- **Stakeholder involvement is essential** when developing a sound system for chemical safety
- **SMEs exhibit utmost reluctance** to communicate risks to the stakeholders
- Developing **suitable mechanisms for stakeholder involvement with calculated risk** is the challenge in improving chemical safety with Responsible production approach
- **Both Top down and Bottom up approaches are required** to take chemical safety initiatives forward in SMEs

1/13/2015 36

Thank you!

1/13/2015 37

EHS Engaged Research

ENOC Group Environment, Health, Safety & Quality Compliance

ENOC

From High Risk – To High Reliability Through Effective EHS Governance

CCPS – Global Summit on Process Safety

15-16th December 2014, Mumbai, India

Presented by: Waddah S. Ghanem Al Hashemi
BEng (Hons), DipEM, DipSM, MSc, MBA, FEI, AFIChemE
Director, EHSQ Compliance

Introduction

ENOC

- Study Value – Objectives
- Developing the Model
- Conceptual Model & Earlier Work
- Defining HROs - Attributes
- Key Themes
- Emerging Model
- Key Findings – Focus and Board Structures
- Conclusions

Study Purpose & Value

- **Purpose:** Major Study exploring a conceptual framework for an effective Safety/EHS Leadership Model for Company Board of Directors
- **Originality/Value:** Very few studies have been undertaken looking at Senior Leadership's perception of the role of the Board of Directors and Corporate Governance of Companies in the context of Health and Safety Very little/no work has been done on this in the GCC area.
- **Basis:** In-depth Literature Review in many areas within the context of Safety, EHS, Corporate Governance and Leadership – followed by Mixed Methods
- **Scope:** More than 25 major organizations, 30 leaders interviewed
- **Type:** High Risk – High Reliability Organizations.

Objectives

Key Objective: *Gain better understanding of the perceptions and views of senior managers on the role of BoD in providing EHS leadership in organizations.*

Explore:

- Underpinning reasons underlying perceptions and views?
- What are the most important thematic areas which drive EHS leadership?
- What other areas are important to consider and understand (other themes)?
- What are the themes that connect Corporate Governance, EHS/Safety and Leadership?
- Study the Impact of Board Structure on EHS Leadership.

Conceptual Model

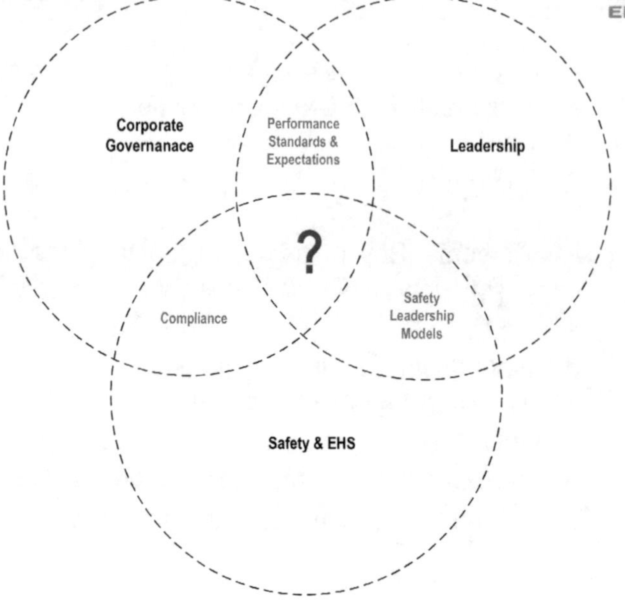

Earlier Research - Board Responsibilities
(Senior Managers – Al Hashemi (2012)

Figure 12: Response to Top-Three Roles and Responsibilities of a Board of Directors from Structured Interviews (COMPANY C) N=18

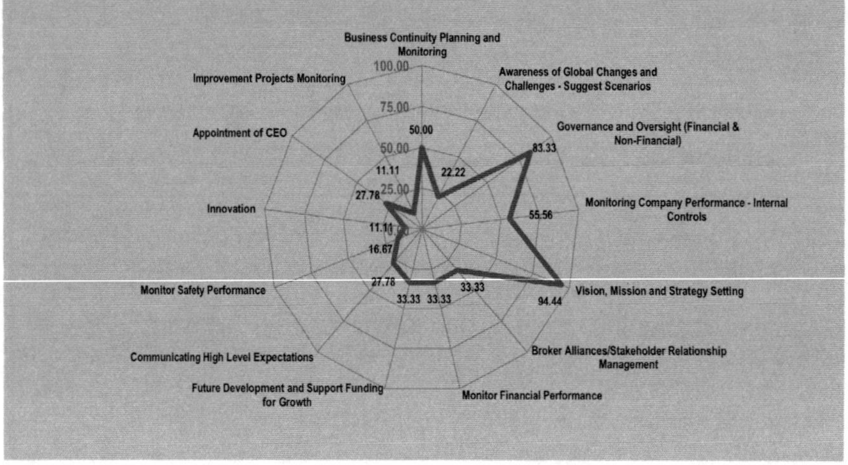

High Reliability Organizations - Definitions

Are organizations that exhibit a ***strong learning orientation; prioritization of safety over other goals; continual training and development****; and an* ***emphasis on checks and maintaining the safety performance****.*

Definition based on HSL (2011); Hopkins (2007) and OECD (2012)

High Reliability Organizations - Attributes

Organizational prioritization of safety and the share performance goals throughout the organization.

No	Characteristic	Implication
1	Dynamic Leadership Shift	Decision making is hierarchical during routine periods with clear responsibilities during emergencies, the organization migrates to a structure in which leverages on the members within the organization who have the expertise.
2	Systematic Intervention	They manage by exception and thus managers focus of strategic and tactical decisions and interfere seldom with operational issues which are delegated and covered by clear processes.
3	Learning Organization	Climate of continuous training and learning. ***Learning organization which uses higher orders of learning to continually improve.***
4	Multi-Communication	Several channels are used to communicate safety critical information – timely communication of information during normal and emergency situations
5	Redundancy	In-built redundancy and the provision of back-up systems in case of a failure. ***"a strategy of redundancy beyond technology".***

Source: Adapted from HSL (2011) & from http://high-reliability.org

Key Themes – From Literature

HSE Knowledge and Competence (of CEOs and Directors)	Can be defined as the basic knowledge of HSE matters to lead as a CEO or direct as a Board Member and this defines the basic competence requirements for this job with respect to HSE.
Safety Leadership	Can be defined as the leading actions driven from a sense of responsibility and accountability for HSE and Safety.
Risk Management	This can be defined as the holistic evaluation, review and monitoring processes for the control of HSE and Safety risks that an organizations may face
Influence and Accountability	Can be defined as the appreciation to the impact of decisions and actions that a senior executive and board member may have and the accountabilities that carries as a consequence.
Developing a Safety Culture and Communication	Can be defined as the development of the organizational culture where safety is a core value and where open, transparent communications are a norm.
Reporting Structure and Hierarchies	Can be defined as the reporting levels and access to senior decision makers. This is for both persons and functions
Legal Imperative for Safety	Can be defined as matters of HSE and Safety which have any legal bearing and impacts.
Operational Excellence & Strong Integrated Management Systems	Can be defined as the systematic management of safety, health, environment, reliability and efficiency to achieve world-class performance.
Monitoring of HSE Performance	Can be defined as the processes of target setting, and monitoring HSE/Safety performance and subsequent management actions to improve and maintain performance.

Model from Desk-Top Review

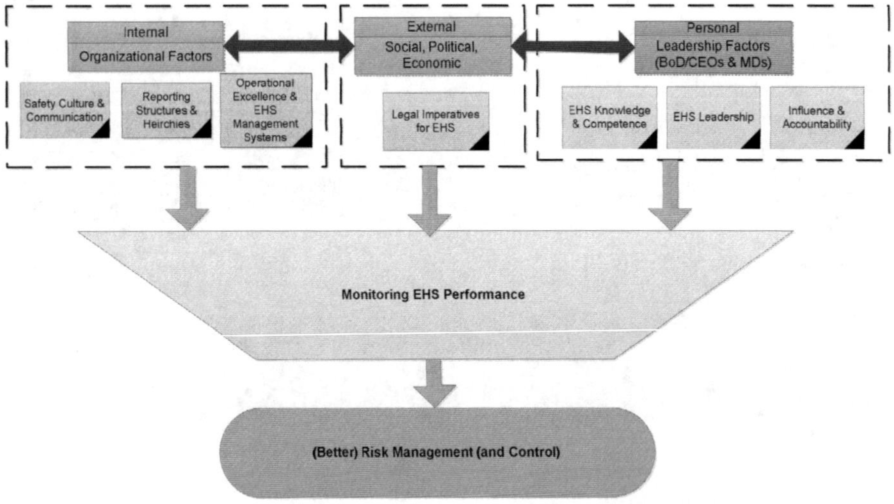

Key Findings – Quantitative Surveys

Descriptive Stats - Top Themes

- Monitoring EHS Performance;
- Influence and Accountability;
- Operational Excellence and EHS Management Systems
- Safety Leadership

Common Area = Monitoring EHS Performance

Key Findings – Quantitative Surveys

Descriptive Stats - Lowest Themes

- Risk Management
- Operational Excellence and EHS Management Systems

Common Area = Risk Management

Note: (*Statistically*) Significant differences in agreement were observed between Oil and Gas (Process) and Non-Oil and Gas (e.g. Aviation, Manufacturing etc)

Key Findings – Impact of Board Structure

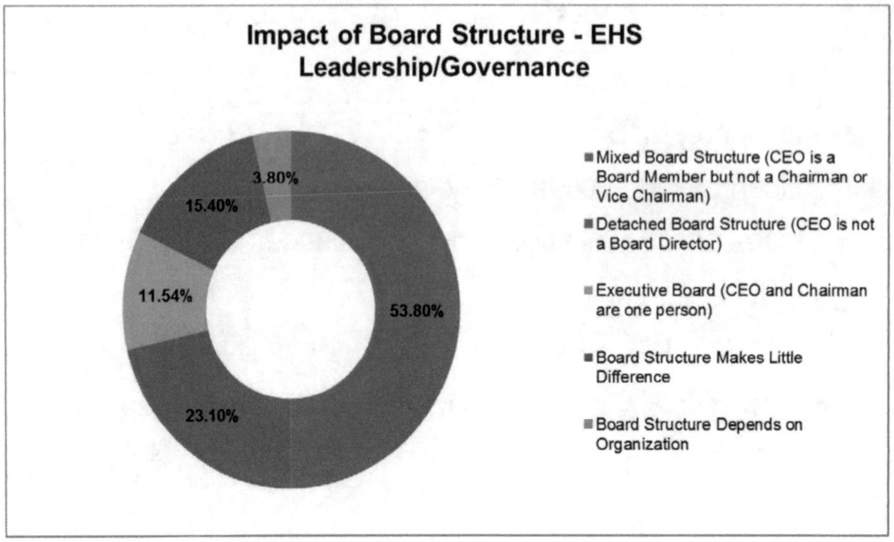

Key Findings – Impact of Board Structure – Reasons Sighted

- ***Mixed Structure***
  - Alignment
  - Collective Efforts
  - Shared decision making process with those who have competence in the EHS issues
  - Understanding of the implications of decisions on Safety
- ***Detached***
  - Oversight / Independence / Separation of Roles
  - Objective Role
  - Clear demarcation of roles, responsibilities and accountabilities
- ***Executive Structure***
  - Full Responsibility & Accountability – leads to better Leadership
  - One-stop-shop
- ***Structure Has no Bearing***
  - Depends on Board make-up of individuals, experiences and general quality of the members of the BoD.
  - Depends on Regulations more than company
  - Depends on organizational maturity.

Key Findings – Emerging Model

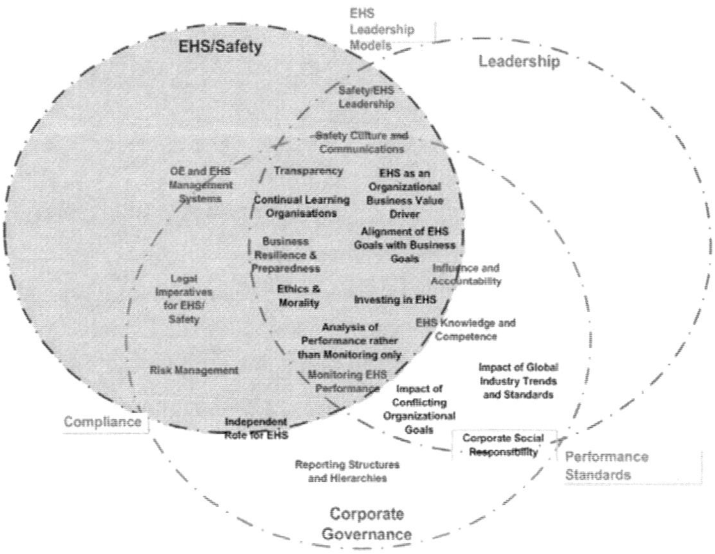

Key Findings – Emerging Model

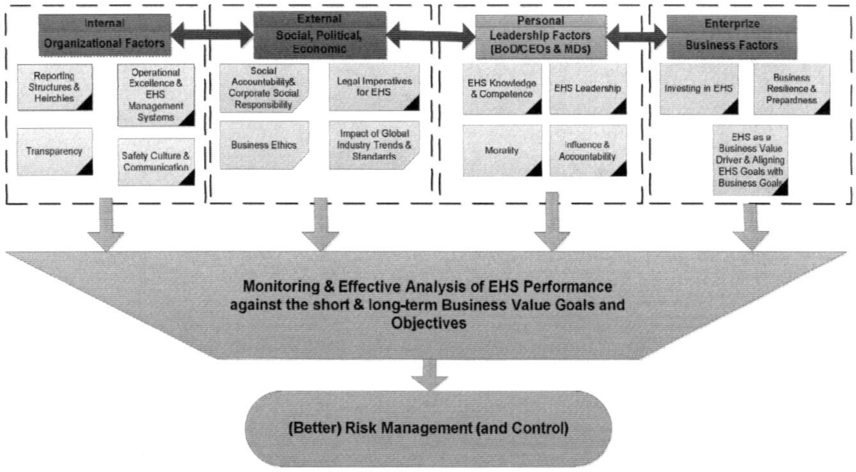

Conclusions

ENOC

(1) **Very little research exists** looking at the important key drivers for **Safety leadership in an Operational Corporate Governance Perspective** especially in GCC

(2) **9 themes** (evolving from a Literature Review) were tested and all seemed to be important although some were more important such as **Monitoring EHS Performance; Influence and accountability Safety Leadership; Safety Cultures and Risk Management.**

(3) Some key differences are there between O&G and Non-O&G HR organizations especially in there approach to **RM; OE/EHSMS, Monitoring EHS Performance and Legal Imperatives for Safety.**

(4) **More than 50%** of the leaders interviewed saw that **Mixed Boards** were best for EHS leadership. **Nearly 25%** saw merit in having **detached boards**.

(5) The Model Developed from this research reflects the complexities involved in Leading and Governing EHS to create HROs.

Thank You for your attention.......

Join Us in 2015 - Dubai, United Arab Emirates

ENOC

GLOBAL HSE

Dubai, UAE

GLOBAL HSE CONFERENCE & EXHIBITION

22-26 February, 2015, Dubai World Trade Centre

- *To find out more -*
  http://www.globalhseconference.com

Process Hazard Analysis - Foundation for Managing Operational Risks

Palaniappan Chidambaram, DuPont Sustainable Solutions

INTRODUCTION

Operational risks can be defined as the potential loss due to possible or foreseeable worst case consequences with a realistic probability of occurrence during facility operation, where such risks span across the entire value/supply chain. Typical operational risks include employee and public health and safety, community economic impact, environmental damage, asset damage, business interruption, reputational loss, regulatory compliance and right to operate. These risks are managed effectively in an organization when events leading to such risks are well understood, communicated and acted upon. In order to achieve this, it is essential to recognize risks associated with the process and operations through the lifecycle of the facility and implement systems to manage them effectively.

On the other hand, organizations that understand the importance of process safety management (PSM) do not always realize the benefit of implementing Process Safety Management Systems despite significant effort and leadership commitment. It can be primarily attributed to poor risk governance and risk recognition, not enough differentiation between high and low hazards/risks and a lack of understanding of the inter-relationship between different PSM elements , resulting in significant time and effort spent on issues and concerns that are of less importance. This often leads to loss of momentum in implementation and low morale among the employees involved.

Figure 1 depicts a model to represent operational risk management in a comprehensive manner and is designed to provide organizations with a better understanding and highlight the importance of risk governance and recognition in managing operational risks. This paper will discuss the typical issues in risk governance, benefits of a differentiated hazard based implementation approach, what Process Hazards Analysis is and how it enables management of operational risks effectively.

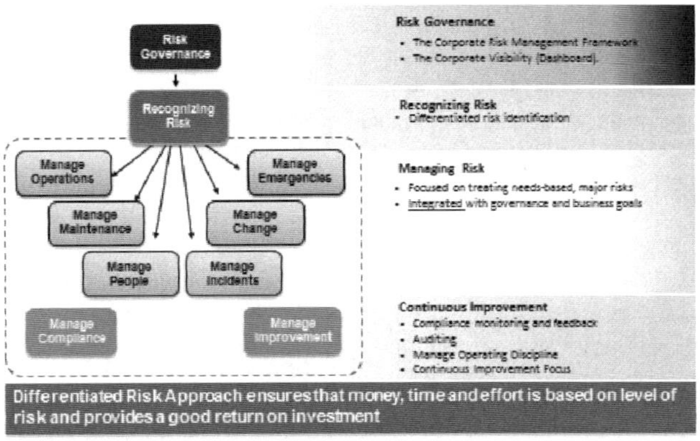

Figure 1 – DuPont Sustainable Solutions' approach to managing Operational Risks

RISK GOVERNANCE

In order to consistently manage risks and make informed decisions, it is essential to have a well codified risk management process and governance structure to execute it. The key element of risk governance is the acceptable risk criteria of the organization which needs to be defined in the risk governance framework. In most cases, risk criteria are represented in the form of a qualitative/semi-quantitative risk matrix. A qualitative or semi-quantitative risk matrix is used during What If/Checklist review, Failures Modes Effects Analysis (FMEA) and Hazard and Operability (HAZOP). Other common areas where risk matrix based decisions are made include prioritization of inspection and maintenance activities, and recommendations from auditing and incident investigations. Based on assessments at several client sites in diverse sectors, we have noticed the following practices with respect to the risk matrix:

1. Use a variety of risk matrices without understanding the underlying risk criteria in the specific risk matrix being used and the acceptable/tolerance risk level of their organization, and
2. A lack of augmenting the risk evaluation with appropriate quantitative studies, especially for high consequence and low frequency events when using qualitative/semi-quantitative risk matrix.

Such practices result in an underestimation of risk, an increase in unmitigated operational risks and a false sense of security at all levels of the organization.

IMPLEMENTATION APPROACH

It is essential to recognize that businesses, sites, facilities, and operations within an organization have a spectrum of hazards ranging from low to very high hazards of concern. Often, equal importance is attributed to all hazards resulting in dilution of effort with minimal reduction of risk. In some cases, more emphasis is placed on work place safety hazards such as slips, trips and falls compared to process safety hazards such fires, explosion, and environmental releases.

In order to be effective in managing operational risks, we need to systematically identify hazards that have the potential to cause catastrophic consequences, differentiate them from those that have minor or negligible consequences, and invest in resources that commensurate the significance of risks. The Operations Risk Management funnel is shown in Figure 2. It is important to develop criteria for categorizing high hazard and low hazard operations and guidance for application of different methodologies, techniques and approaches to assess and manage them. Such a differentiated approach enables prioritization of implementation and realization of benefits from operational risk reduction.

RISK RECOGNITION: PROCESS HAZARDS ANALYSIS

Process Hazard Analysis (PHA) is a framework for identifying the process hazards, hazardous events and their consequences, evaluating the adequacy of the existing safeguards and determining the need for additional safeguards to reduce the risks to an acceptable/tolerable level based on an organization's risk matrix. Some of the common challenges observed in the industry are:

1. Initial PHA done by project team without involving operations resulting in an incomplete assessment
2. Process Technology or Process Safety Information not handed over by the project team to the operations team or the information is not kept up to date
3. HAZOP considered as the beginning and end of PHA
4. PHA outputs not considered in designing the remaining PSM elements. For example, emergency scenarios are not developed based on hazard analysis and the mechanical integrity or asset integrity program is not in line with the operational risks identified

5. PHA not done periodically or when there are changes in process and/or equipment resulting in increase of potential unmitigated risks
6. Lack of Rigor in implementing PHA recommendations
7. PHA conducted without consideration to both major internal and external incidents, and
8. Not considering PHA in the early stages of project design.

Figure 2 – Operations Risk Management Funnel

PROCESS HAZARDS ANALYSIS ≠ HAZOP

Process Hazard Analysis is often equated to HAZOP, but PHA is much more than a HAZOP study. Process Hazard Analysis is a structured framework which includes:

1. Preparation and planning to ensure right resources are identified and up to date information is available for analysis
2. Hazard identification
3. Process Hazard Evaluation using different techniques (What-If/Checklist, FMEA, HAZOP)
4. Consequence Analysis to determine the onsite and offsite impact of hazardous event
5. Facility Siting Analysis to ensure occupied buildings and buildings housing critical functions are adequately protected
6. Interlock Evaluation using Layers of Protection Analysis (LOPA) to determine adequacy of safeguards and need for additional safeguards for high consequence/high risk events; and
7. Human factors checklist review
8. Inherent safer process checklist review
9. Risk assessment and recommendations management
10. Detailed Documentation of entire PHA process.

As discussed in previous sections, it is essential to apply an appropriate combination of qualitative and quantitative techniques that commensurate the degree of risk associated with the unit, process or scenario as well as to overcome the limitations of qualitative studies. Figure 3 depicts the PHA framework and shows key outputs that form a basis for managing operations, emergencies, incidents, maintenance, and training.

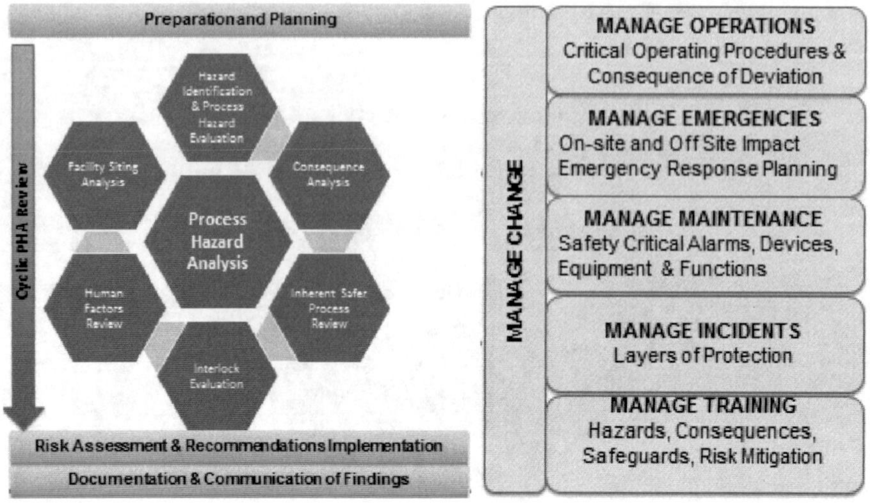

Figure 3 – Importance of Process Hazards Analysis

RISK REGISTER AND COMMUNICATION

Upon completion of PHA, a register of high risk events is essential for focused operational risk management and to help ensure risk levels are sustained or reduced further during the lifecycle. The importance of communication of high risk and worst consequence events to employees involved in operations and maintenance of facility is often overlooked. As a result, employees are not aware of the risks and the importance of maintaining the safety critical devices, alarms, equipment and functions for instance.

Palaniappan Chidambaram is the Global Solutions Architect for Process Safety Management with DuPont Sustainable Solution, based in Singapore.

For further details on the entire implementation approach, join us on
15th December 2014 at the CCPS Global Summit on Process Safety during the session
Disciplined Adherence to Standards to hear more from Mr.Palaniappan Chidambaram

You can also contact:
Ravindran Menon - Ravi.Menon@dupont.com
Deepti Rajendran Venkateswar - Deepti.Venkateswar@dupont.com

RELIABILITY ISSUES IN PIPELINES A CASE STUDY

B.K.GUPTA
Addl. Director
Oil Industry Safety Directorate

Structure of Presentation

- Backdrop
- Reliability & Safety
- Case Study
- Summary

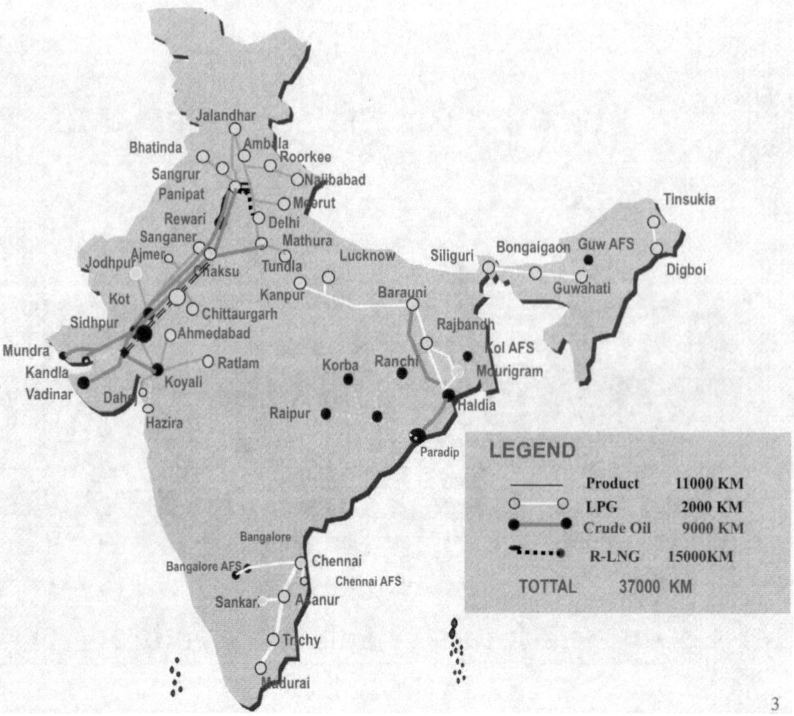

3

Cross Country Pipeline System

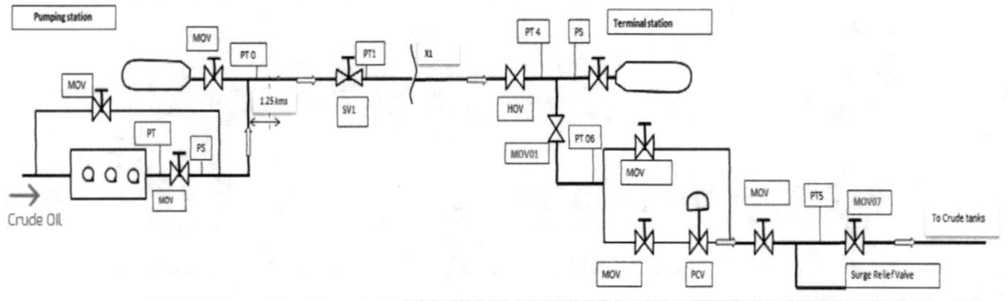

Leak site located in a remote area with difficult terrain.

View of Challenging Gradient w.r.t. Pipeline Maintenance

A CRACK AT SITE OF DENT/BUCKLE.

RELIABILITY & SAFETY CAN'T BE SEPARATED

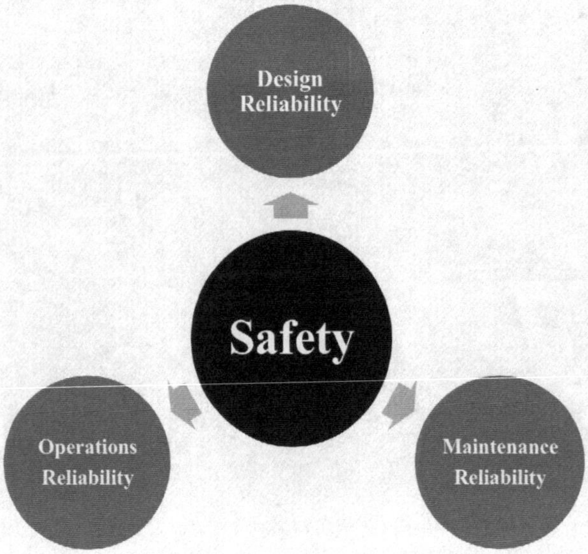

8

DESIGN RELIABILITY

Design reliability achieved by :

1. Compliance with the appropriate standards (API, ASME & OISD).
2. Selection of the grade of pipe material and wall thickness of pipe, type of valves and other pipe fittings.
3. Various factors like internal pressure, static loadings and weight effects of the pipe etc. are taken into consideration throgh STD.
4. Identification of high consequence areas & location class.
5. Special protection at crossings like river/canal, rail, road etc.,
6. HAZOP and Risk Analysis should be done at design stage

9

DESIGN RELIABILITY

Design reliability achieved by :

7. Location of "Sectionalizing valves" is important.
8. Sectionalizing valves should be MOV (Motor Operated Valve)
9. Review in P&ID for ensuring feature of de-pressurization and station bypassing.
10. Shutdown & isolation logic in control system.
11. Provision of SCADA systems for continuous monitoring and control of remotely located Sectionalizing Valve.
12. Leak detection system with provision of identification of leak location.
13. External coating & cathodic protection for external corrosion control.
14. Inhibitor injection for internal corrosion control case.

10

OPERATION RELIABILITY

Operation reliability achieved by :

1. Comply with standard operating procedures (SOP).
2. Pressure set points etc. should be appropriately defined in the operation philosophy.
3. ESD logic (cause and effect matrix) should be available in control room.
4. Pigging has to be carried out as per guidelines
5. The results of the pig-residue analysis should be analysed.
6. Safety System should not be bypassed.
7. Ensure dedicated , reliable and standby communication system (OFC, leased line etc.)
8. Ensure reliable functioning of SCADA system.

11

OPERATION RELIABILITY

Operation Reliability Achieved By :

9. Hand operated valves (HOV) installed on old pipelines should be converted into motor operated valve (MOV) for remote operation and prompt action for sectionalisation of Pipeline.
10. Proper analysis of cathodic protection monitoring reports.
11. Any change w.r.t. Basic design conditions should be done after proper study/ analysis and through a proper "Management Of Change" procedure
12. **Human Reliability:**

    a. Ensure availability of Operation manual

    b. Deploy adequately trained and skilled manpower for pipeline operations.

    c. Training on auxiliary system like CP, SCADA, ESD, Earthing and FF equipments etc.

    d. Training on "How to handle Emergency situation."

    c. Motivation, recognition and skill development of manpower should be looked into

12

MAINTENANCE CONSIDERATIONS

1. Aim is to minimize breakdown maintenance
2. More focus to be given on predictive & preventive maintenance.
3. Condition based monitoring should be encouraged. Operators must be sensitive to understand the language of the machine.
4. All routine inspection results (CP survey, Coating survey, Soil Resistivity, IPS, Pig residue analysis etc.) should be analyzed and accordingly action should be planned.

13

MAINTENANCE CONSIDERATIONS

Pigging is a very important activity in pipeline

Pipeline Pigging - Pigs are inserted into the pipeline while it is operational and are carried along by the fluid being pumped.

- Appropriate pigs to be used (foam pigs instead of scrapper pigs is not proper).

Examples of the Types of Pigs in Use Today

CLEANING PIGS

14

MAINTENANCE CONSIDERATIONS

1. CLEANING PIG - Provide a way to clean debris and scale from the inside of the pipe,
2. INTELLIGENT PIG - Inspect or identify type of defects like metal loss, dent and crack etc. of the pipe.

INTELLIGENT PIG

15

MAINTENANCE CONSIDERATIONS

Pipeline coating & cathodic protection

Protective Coating : To prevent direct contact between the Pipe and the environment, to isolate the metallic pipe and to protect from corrosion.

Coating Surveys like DCVG/ Pearson survey is carried out to identify coating defects.

Coating Newly Installed Pipe for Corrosion Control

16

CASE STUDIES-OISD

CATASTROPHIC FAILURE DUE TO POOR MAINTAINANCE

NATURAL GAS PIPELINE

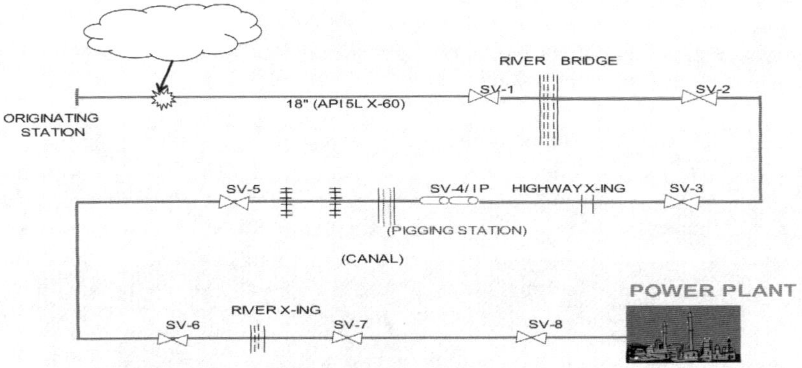

Length of pipeline	**Approx 200 KM**
Dia of Pipeline	**18" OD**
Flow:	**0.7MMSCMD**
Pressure:	**42 Kg/cm2**

CATASTROPHIC FAILURE DUE TO POOR MAINTAINANCE

Background

- Cross country pipeline – supplying Natural Gas to power plant.
- The pipeline was being operated at a lower flow rate.
- Pipeline was transporting wet gas.
- The gas contained CO_2 and some traces of Sulphur.
- Leakage took place at the lowest point where the pipeline was crossing a canal at 6 o'clock position.
- There has been a history of previous leaks near the place of accident
- Leaks were repaired by clamping.
- After repairing damaged coating was not repaired.

CATASTROPHIC FAILURE DUE TO POOR MAINTAINANCE

Background

- Only foam pig was used for cleaning.
- It was raining on the day of the incident, so we inferred that vapour cloud of highly flammable hydrocarbon could not disperse due to heavy weather.
- A tea stall owner. who turned on the stove in the morning, possibly provided the source of ignition.
- **A major explosion occurred in Pipeline gave multiple causalities.**

CATASTROPHIC FAILURE DUE TO POOR MAINTAINANCE

- Fire extinguishing took 2 hours.
- There was a long delay in closing the nearby sectionalising valve (SV). If remote operation of SV was functional, the extent of damage could have been reduced.

Root Cause of Incident

- Leak due to internal corrosion because of presence of condensate. Presence of water and CO2 form weak Carbonic acid, which induce internal corrosion.
- Improper pigging – foam pig used instead of scrapper pigs.
- Pig residue analysis not carried out.
- Use of wet natural gas instead of dry natural gas.
- Not using corrosion inhibitor.

CATASTROPHIC FAILURE DUE TO POOR MAINTAINANCE

Recommendations and Lessons Learnt

- Proper cleaning pigging to be carried out regularly followed by pig residue analysis.
- Instrumented pig survey to be carried out periodically and the anomaly rectification to be carried out as per the recommendations .
- Dehydration facility must be provided for removal of water and condensate prior to feeding natural gas in the pipeline.
- Corrosion inhibitor to be used in the line as per requirement.
- Proper maintenance and inspection practice to be followed.
- External Coating to be repaired promptly
- In addition to OFC communication there should be a leased line as back-up.
- Remote operation of sectionaling valve to be restored.

EXAMPLES OF POOR MAINTAINANCE

Damaged condition of coating

Condensate in field after leak

Summary

- MAINENANCE ACTIVITIES AND SCHEDULE AS GIVEN IN STANDARDS MUST BE ADHERED TO.
- INSPECTION RESULTS SHOULD BE PROPERLY ANALYSED AND RECOMMENDATIONS ADHERED TO.
- FOCUS SHOULD NOT BE LIMITED TO DESIGN ASPECT ONLY BUT ALSO ON OPERATION AND MAINTENANCE RELIABILITY.

THANK YOU

A risk evaluation method in Mitsubishi Chemical Group

Masaki Nakagawa
Process Safety Technology Department,
Mitsubishi Chemical Corporation
1-1, Marunouchi 1-chome, Chiyoda-ku, Tokyo 100-8251, Japan
2109797@cc.m-kagaku.co.jp
Keywords: HAZchart, Fault tree, PHA_Organizer

Abstract

Recently, the number of accidents which occur owing to the plant staff's deficient knowledge or the shortage of experience is increasing because plant staff's opportunity to experience troubles is decreasing and veteran operator's retirement increases. Therefore, it is necessary to make a mechanism of sharing the knowledge and the perspective from veteran staffs to young staffs.

One of the most successful risk evaluation methods in Mitsubishi Chemical Group is HAZchart analysis, which can carry out easily from development of an accident scenario to quantitative evaluation. In modeling accident scenarios, HAZchart analysis clearly distinguishes success of recovery action from failure of recovery action, by using symbols. Therefore, according to the procedure of HAZchart analysis, worst case scenarios are able to be developed easily.

HAZchart analysis method was packaged as a computer program called PHA_Organizer. PHA_Organizer has some risk analysis support functions to facilitate sharing a mutual thinking process and knowledge transfer. It is expected that these functions contribute to improve plant staff's understanding and awareness for chemical plant safety.

1. Introduction

Recently, the number of accidents which occur owing to plant staff's deficient knowledge is increasing because plant staff's opportunity to experience troubles is decreasing. Steady operation in a plant makes plant staffs unable to imagine abnormal situations of a plant. When performing risk evaluation, plant staffs do not regard the result of risk evaluation as a realistic situation due to poor understanding of their plant. Therefore, tools which improve plant staff's understanding and awareness for chemical plant safety are called for.

2. HAZchart analysis

a) Procedure for HAZchart

One of the risk evaluation methods used for safety assessment and safety review activity in Mitsubishi Chemical is HAZchart analysis. HAZchart analysis was developed by Mitsubishi Chemical Corporation and Mitsubishi Research Institute in 1996. It can carry out easily from development of an accident scenario to quantitative result.

HAZchart analysis standardizes procedure of accident scenario modeling. Accident scenario is modeled by querying three situations repeatedly. After that, a HAZchart is transferred to Fault Tree (FT) and the probability of accident can be calculated easily.

HAZchart analysis enables us to consider the worst case scenario. This method has been used to check a new process safety level and a modified process safety level. In addition, it has been used to study an existing plant safety level and to train plant staffs to upgrade safety knowledge of their plant.

The procedure for HAZchart analysis is described in the following example (Figure 1). Material A, B and catalyst C are fed in a certain proportion into the reactor. If the ration of the materials were changed, runaway reaction would occur.

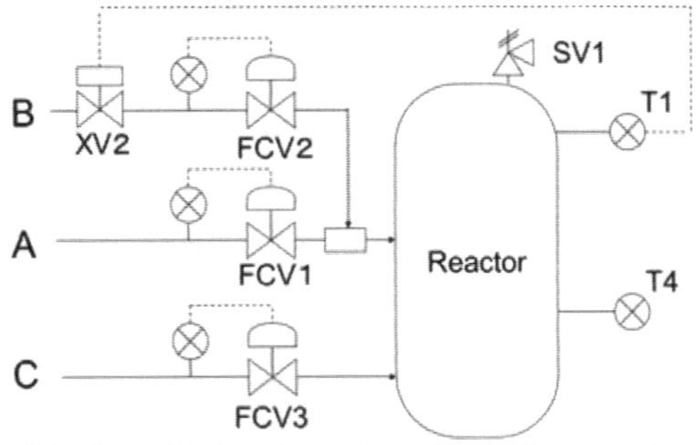

Runaway reaction occurs
in case of (material B) / (material A)> α

Figure 1 Example Flow

First of all, one of the equipment which can potentially influence the process fluid is selected as a starting point for HAZchart. Then the fault modes are assumed to them. In this case, FCV2 which controls the mass flow of material B is chosen. Then the fault mode is assumed as control valve excessive opening. After setting the fault mode, by querying three situations the accident scenario is modeled (Figure 2). The HAZchart is modeled by querying three situations which are "what is the method of the finding phenomenon ?", "what is the recovery action ?" and "what is the consecutive phenomena ?". The method of the finding phenomenon indicates alarm etc. The recovery action indicates system action or operator action. The consecutive phenomena indicate the result of the recovery action. Each query corresponds to each symbol. The method of the finding phenomenon corresponds to a straight line, the recovery action corresponds to a diamond and the consecutive phenomena correspond to a rectangle. The HAZchart that forms the scenario of an accident is modeled by repeatedly applying these three queries. Figure 3 shows HAZchart of FCV2 excessive opening.

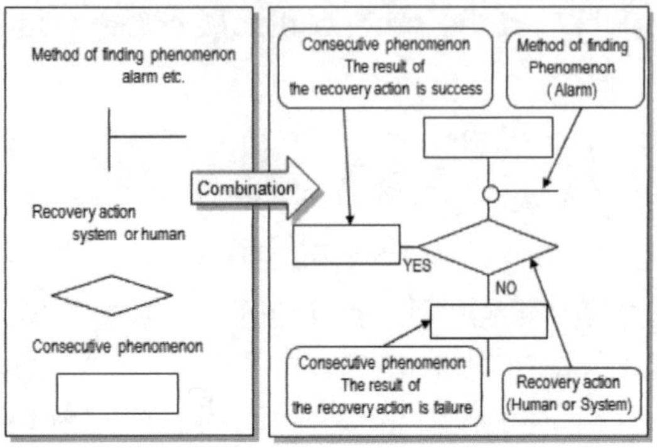

Figure 2 Three queries

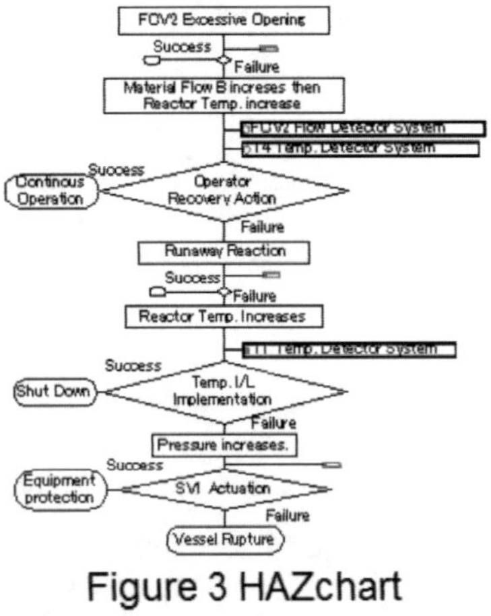

Figure 3 HAZchart

Figure 3 shows HAZchart of FCV2 excessive opening.

Next step is to transfer HAZchart to FT. "Methods of finding phenomenon", "recovery actions" and "consecutive phenomena" in the HAZchart correspond to FT logic. Thus, it is easy to transfer information from a HAZchart to a FT.

The FT corresponding to the HAZchart can be obtained by the following five steps.

1) HAZchart is turned upside down.

2) Rectangular parts which indicate "consecutive phenomena" are arranged on the left side of the FT. Diamond parts which indicate "recovery actions" are

arranged on the right side of the FT.

3) The rectangular parts are combined with the diamond parts by " AND" symbol (Figure 4).

4) After the transformation, branches of FT are divided up to a level that has equipment reliability data or human error data.

5) After the equipment reliability data are set up in the FT, the probability of top event can be calculated (Figure 5).

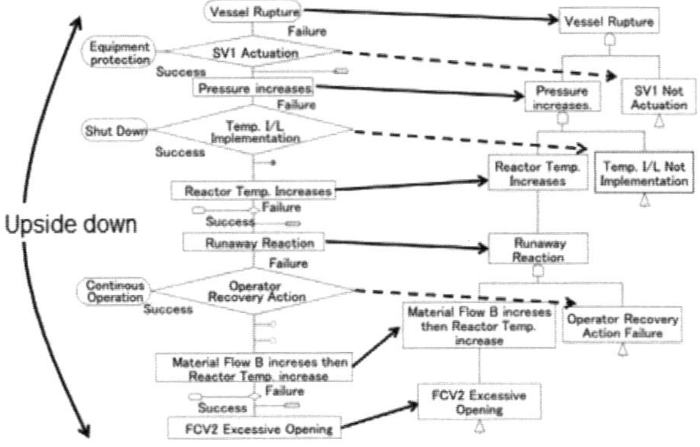

Figure 4 Transformation from the HAZchart to the FT

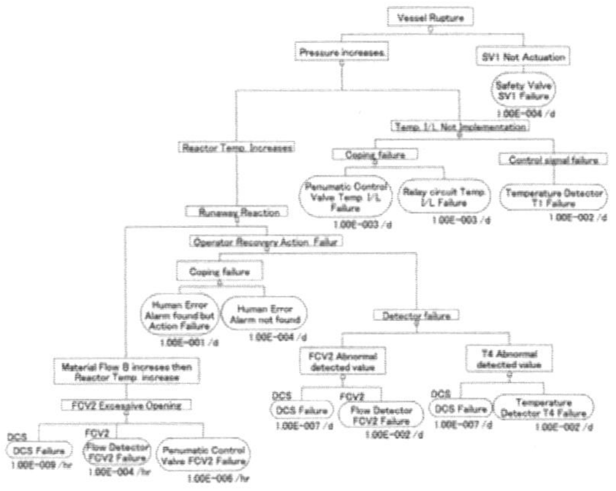

Figure 5 FT from HAZchart

The result of the HAZchart analysis is plotted in the risk matrix. Horizontal line means probability of accident and vertical line means accident levels. Red area and yellow area indicate unacceptable levels and White area and green area indicate acceptable levels. The star symbol indicates that the explosion due to the failure of FCV2 occurs at low probability (Figure 6).

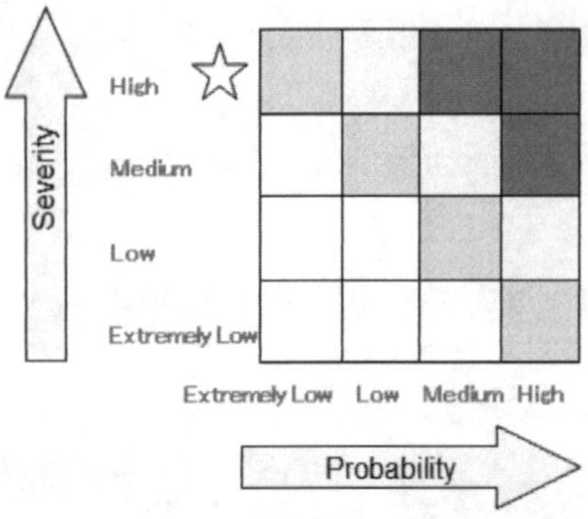

Figure6 Risk Matrix

b) Correspondence of protection layers with HAZchart.

The safety system is designed as a series of protection layers. So, it is natural for plant staffs to think that if a malfunction on a protection layer occurs the situation becomes worse. HAZchart was developed along the same concept of layers of protection. Each protection layer corresponds to each recovery action of the diamond. Figure 7 shows relationship between protection layers and HAZchart.

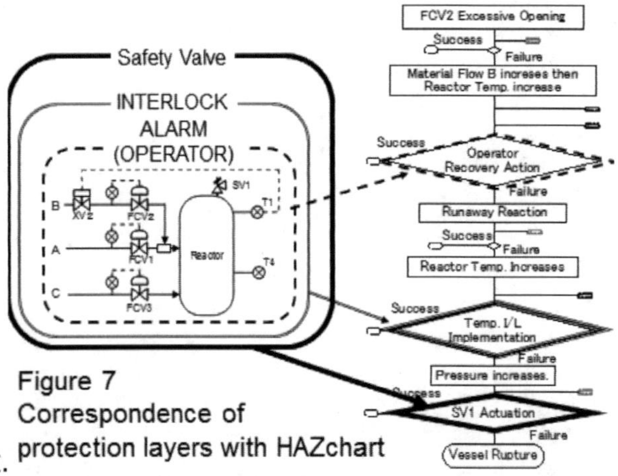

Figure 7
Correspondence of protection layers with HAZchart

c) Development of accident scenario

Performing "what-if analysis", worst case scenarios are sometimes missed because accident scenarios are modeled based on the assumption of safeguard operating. Figure 8 shows a part of result of what-if analysis. The problem of this result is improved by using HAZchart analysis method. In the result of what-if analysis, consequence of FCV2 excessive opening is shut down by the Interlock system. The worst case scenario was missed in the result of what-if analysis because the accident scenario was modeled based on the assumption of safeguard operating. But normally, the accident scenario should be modeled on the assumption of failure of active protection system. In modelling accident scenario, HAZchart analysis clearly distinguishes success of recovery action from failure of recovery action, by using symbols. According to the procedure of HAZchart method, worst case scenarios are able to be modeled easily.

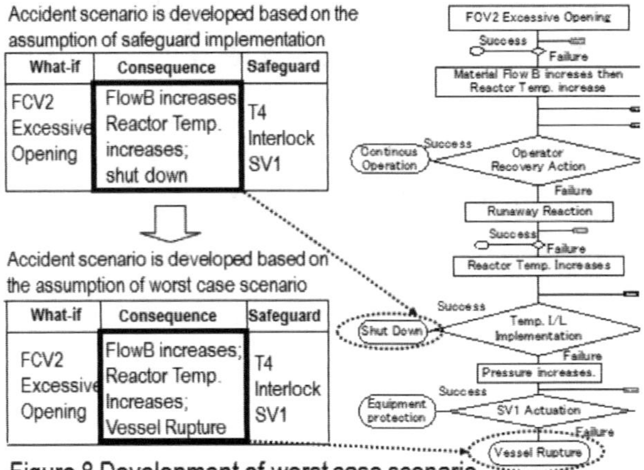

Figure 8 Development of worst case scenario

3. PHA_Organizer

HAZchart analysis method was packaged as a computer program which was called PHA_Organizer. PHA_Organizer has a database which has equipment reliability data and branches of FT. Therefore, after modeling HAZchart, the probability of accident scenario can be calculated automatically. This software was privately used at first. But it became clear that PHA_Organizer could facilitate sharing a mutual thinking process and promoted discussion at a meeting. Consequently PHA_Organizer was significantly improved to make the tool even easier to use and anyone could easily conduct risk evaluations at a meeting. This program is sold by Mitsubishi Chemical group company. The features of PHA_Organizer are described as follows.

a) Minimal cut set display function

A cut set is a combination of basic phenomena which generates the top phenomenon of FT. A minimal cut set is made required to generate a top phenomenon of a system, and is sufficient cut set. The minimal cut set was often expressed as a set. But this expression form is not so easy to understand intuitively. So the function of displaying minimal cut set relationship on the FT has been developed. By this function, it is easy to understand which basic phenomenon lowers the probability of the top phenomenon occurrence intuitively.

b) Common cause specific function

Common cause is an important factor in the FT calculation. The reason why is that the accident would be underestimated if common cause was not set up by mistake to calculate the probability of an accident. To solve this problem, PHA_Organizer could calculate the probability of FT that included common cause by recognizing equipment as the same equipment if their information was the same.

c) Redundant system support function

A redundant system is adopted as one of the means of increasing reliability of an interlock system. It is troublesome to describe a redundant system in a FT. Therefore, the support system that can describe a redundant system easily has been developed.

4. Case study by PHA_Organizer

a) Original flow

Figure 9 shows an original flow of case study. Mixture of liquid and gas is fed into a separator and separated into liquid and gas. The LCV1 controls to maintain liquid level for preventing gas leakage into the liquid line. The starting point is LCV1 excessive opening. The accident scenario would occur in order of 1 to 4 in Figure 9. The diamond in the HAZchart shows there is one protection layer in this accident scenario.

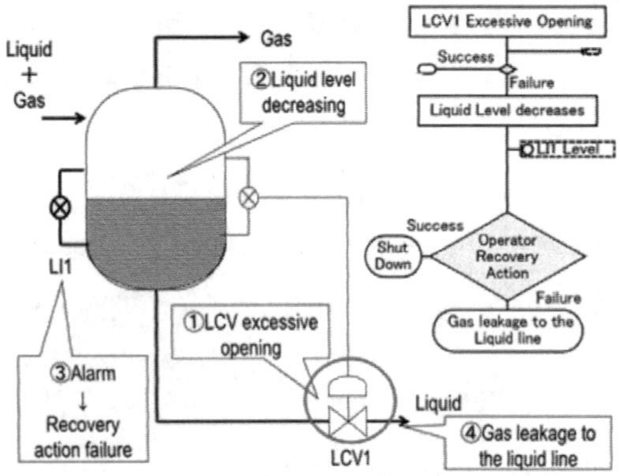

Figure 9 Original Flow

Figure 10 shows FT of the original flow and Figure 11 shows the result of the calculation. The result of the HAZchart analysis is plotted in the risk matrix. The result of LCV1 excessive opening is in the red area. That means leakage to the liquid line due to the failure of LCV1 is unacceptable.

b) Modified flow 1 (Case study by minimal cut set display function)

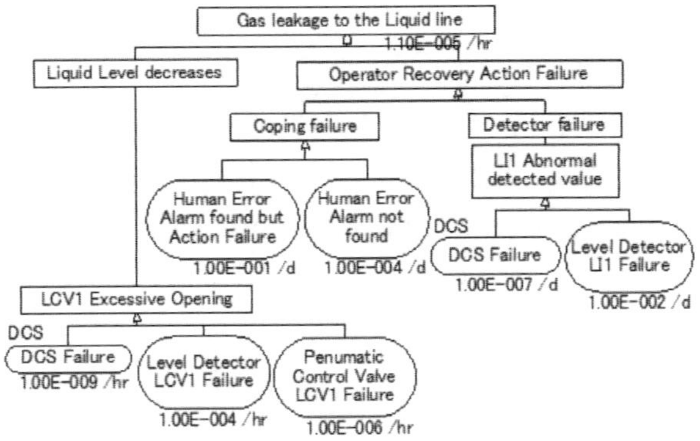

Figure 10 FT of original flow

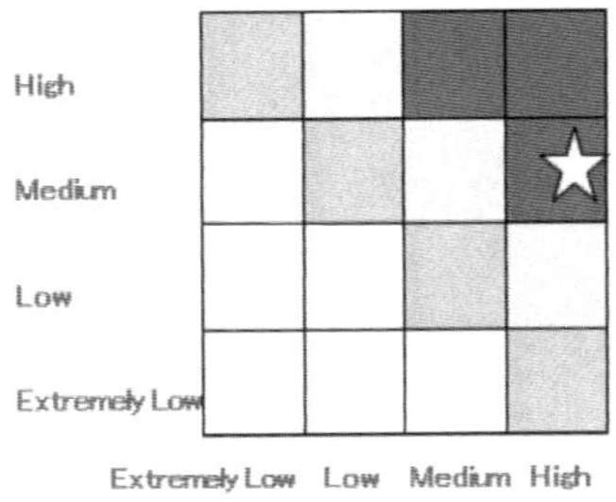

Figure 11 Result of the original flow in the risk Matrix

Figure 12 shows modified flow 1. Two level detectors are added to this flow. In this case, if the liquid level decreased, operator would take a recovery action to prevent from gas leakage by three level alarms. The probability of gas leakage in the case of modified flow 1 is the same as in the original flow case. This result shows the countermeasures of three level detectors are not effective for decreasing the probability of gas leakage.

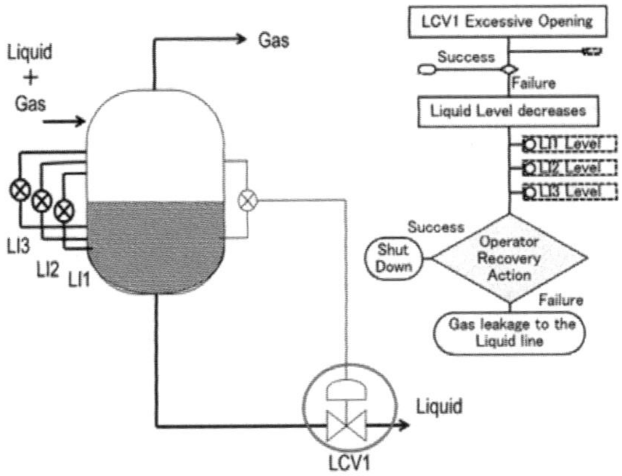

Figure 12 Modified flow 1

Figure 13 shows FT of the modified flow 1 on the PHA_Organizer screen, in which minimal cut set display function is operating. The combination of minimal cut set can be displayed on the FT by changing related line's color and thickness in PHA_Organizer. The highest demand rate of the basic event in the branch of “operator recovery action failure” is “Human error alarm found but action failure”. So, the minimal cut set of the highest probability is the combination of "Level Detector LCV1 Failure" and "Alarm found but Action Failure". That means that even if the number of alarms were increased, the probability of top event could not be reduced by them. The demand rate of "Alarm found but action failure" cannot be changed in this analysis. Therefore, the probability of top event can be reduced by adding other safety systems.

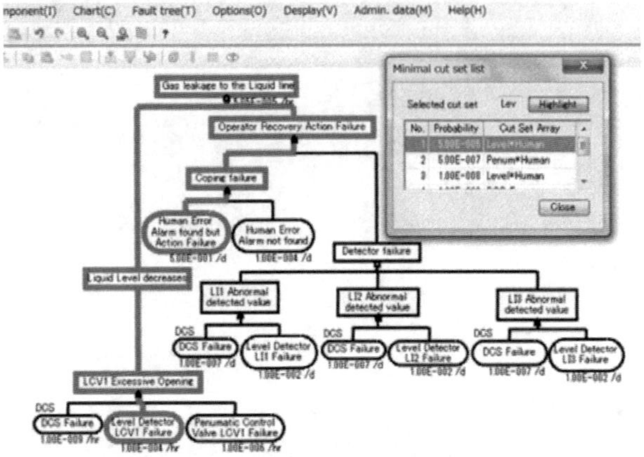

Figure 13 Minimal cut set display function

c) Modified flow 2 (Case study by common cause specific function)

Figure 14 shows case study of modified flow 2. The interlock system is added to the flow. In this case, if the liquid level decreased, the interlock system would operate to prevent from gas leakage. The alarm and the interlock system are protection layers. But the level detector of LCV1 is shared with the interlock system and the level alarm.

Figure 15 shows the FT of modified flow 2. The FT has branches of operator recovery action failure and Interlock system failure. There are the same equipment in the left side and right side in the FT, which are level detector and DCS. They are common cause failure.

Figure 16 shows the FT displayed by minimal cut set function. The minimal cut set with the highest probability is the combination of "Level Detector LCV1 Failure" and "Alarm found but Action Failure". The probability of gas leakage of modified flow 2 is the same as the original flow and the modified flow 1.

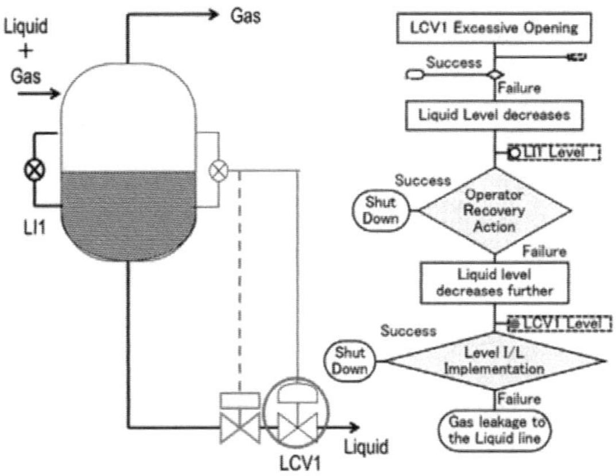

Figure 14 Modified flow 2

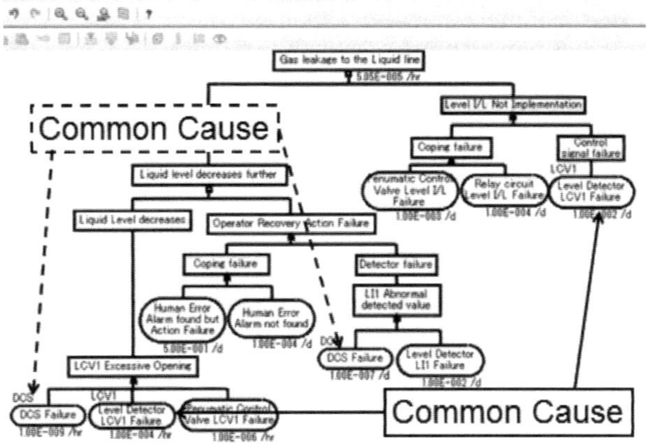

Figure 15 Common Cause in the FT

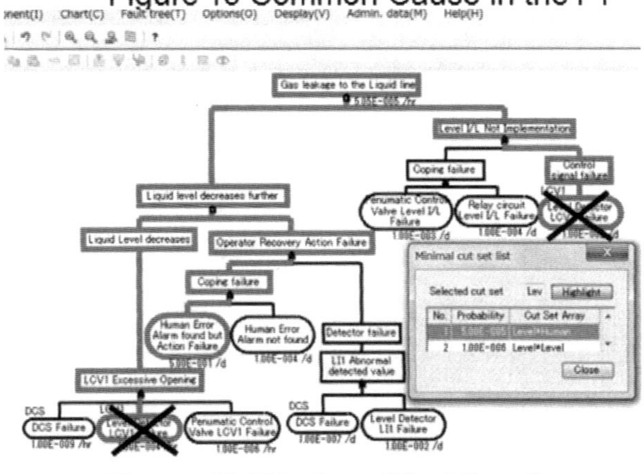

Figure 16 FT of modified flow 2

d) Modified flow 3 (Case study by redundant system support function)

Figure 17 shows modified flow 3. An independent interlock system and an alarm are added to this flow. If the liquid level decreased, the alarm and the interlock system would operate to prevent from gas leakage. This time, the interlock system and the alarm has different level detector. In the event that the level detector failed and the operators did not take the appropriate recovery action, the interlock system would prevent the liquid level from decreasing.

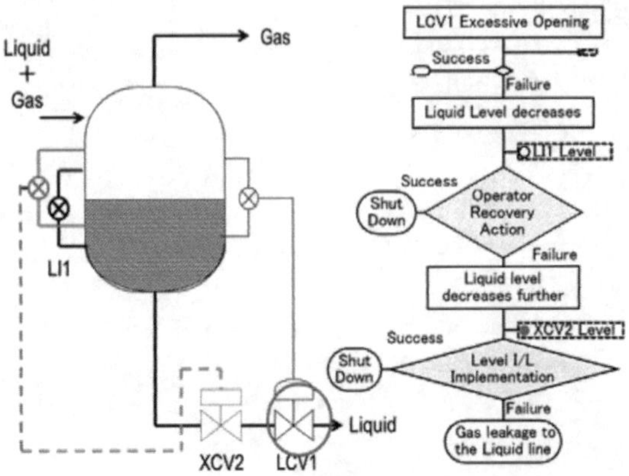

Figure 17 Modified flow 3

Figure 18 shows the probability of gas leakage in this case can be reduced because the interlock system does not share common equipment.

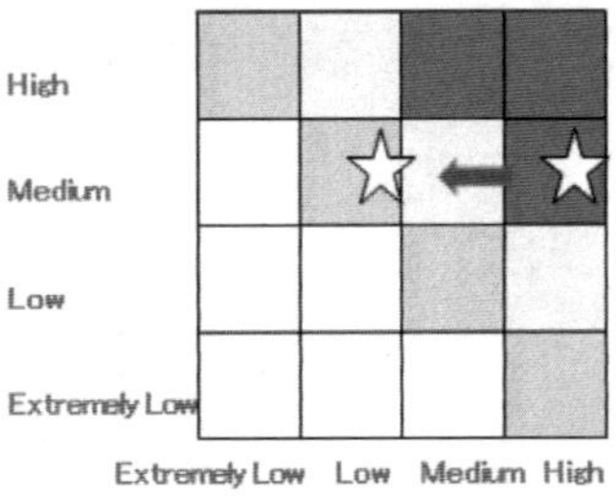

Figure 18 Risk Matrix

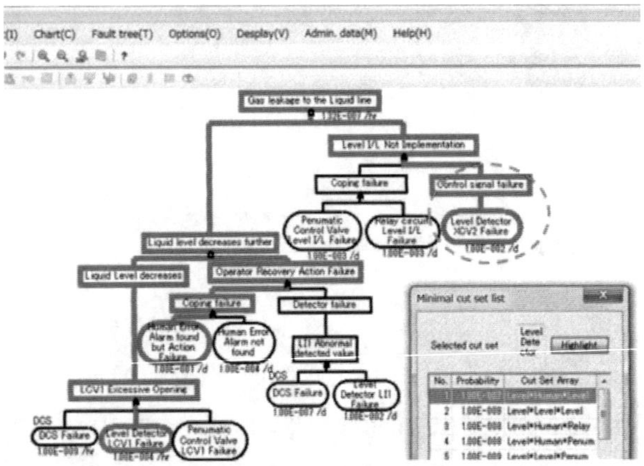

Figure 19 FT of modified case 3

Figure 19 shows the FT of modified flow 3 displayed by minimal cut set function. The highest probability of the minimal cut set is the combination of "Level Detector LCV1 Failure", "Alarm found but action Failure" and "Level detector

XCV2 failure". The demand rate of "Alarm found but action failure" cannot be changed in this analysis. Therefore, the probability of top event can be reduced by improvement of "Level detector XCV2".

Redundant system is adopted as one of the means of increasing reliability of an interlock system. To describe redundant system easily in the FT, the redundant system support system was developed. The procedure for the support system is as follows.

(1) All detectors that are included in redundant system are described in the HAZchart.

(2) All of them are selected.

(3) They are set as a redundant system (m out of n system).

(4) Redundant system of FT is modeled automatically(Figure 21).

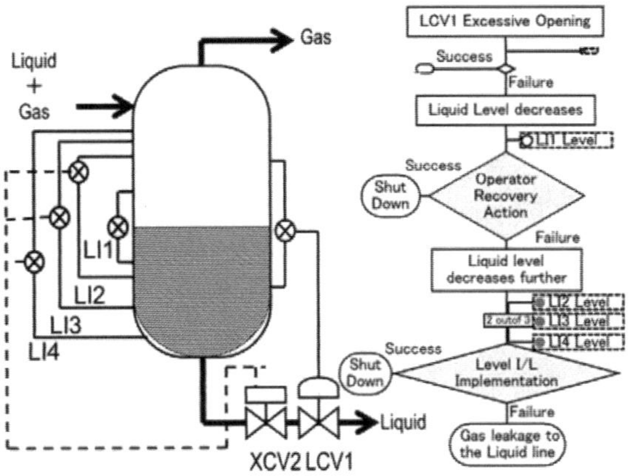

Figure 20 modified case flow

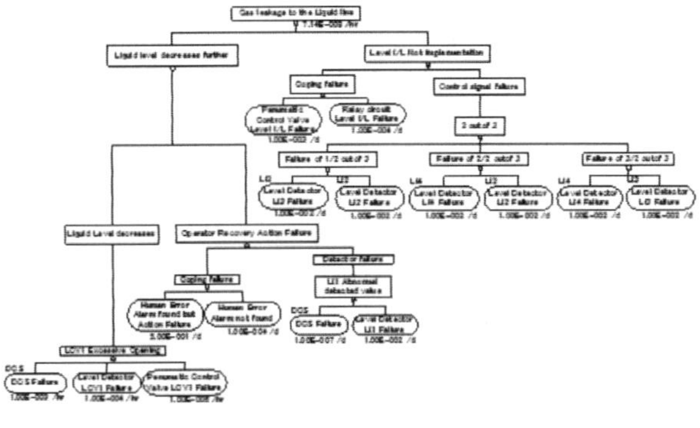

Figure 21 Redundant system in the FT

5. Application

When performing safety evaluation at a meeting, we use two projectors, one set is used for the screen of PHA_Organizer and another set is used for the screen of flow sheets. PHA_Organizer shows thinking process that can facilitate sharing a mutual thinking process. Showing modeling process of accident scenario by PHA_Organizer induces various questions. Plant staff can understand their plant better by answering questions quantitatively through HAZchart analysis. PHA_Organizer is expected as a tool for providing a field which promotes discussion.

Figure 22 shows the evaluation result of an existing process. The results of this study are safe for serious accident. But further study is needed for stable operation

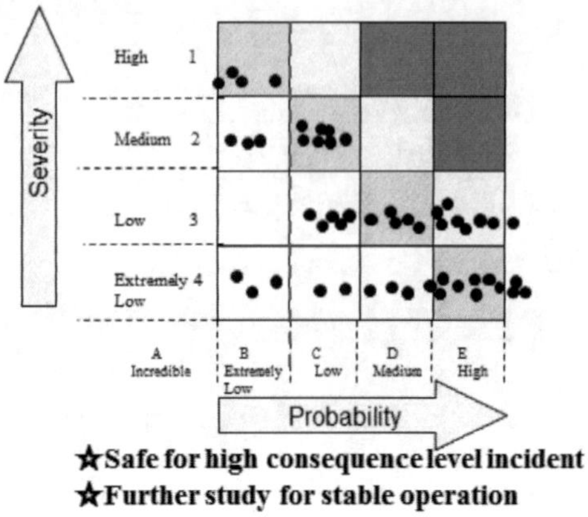

Figure 22 Evaluation result of an existing plant

Figure 23 shows the evaluation result of modified process. By modification of process line and addition of Interlock system, these two points are transferred from unacceptable area to acceptable area.

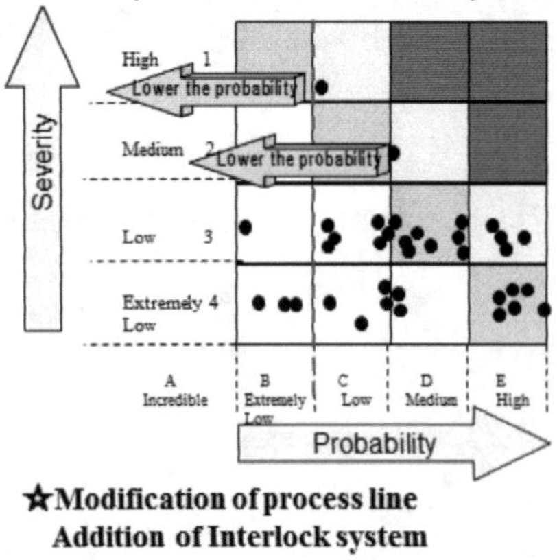

Figure 23 Evaluation result of an modified process

6. Conclusion

HAZchart analysis is a tool for identification of latent hazards. It can carry out easily from development of an accident scenario to quantitative evaluation.

HAZchart analysis can facilitate sharing a mutual thinking process and promote discussion at a meeting.

PHA_Organizer has specific functions which are minimal cut set display function, common cause specific function and redundant system support function. It is expected that these features contribute to improvement of plant staff's understanding and awareness for chemical plant safety.

REFERENCES

[1] M. Nakagawa., Y. Iizuka, Introduction and application of quantitative process hazard analysis – HAZchart, 11th Intl. Symposium of Loss Prevention and Safety Promotion in the Process Industries, 31 May – 3 June 2004, Praha, Czech Republic, 4388-4394.

[2] M. Nakagawa, Y. Mizuta, Development of a worst case scenario, 13th Intl. Symposium of Loss Prevention and Safety Promotion in the Process Industries, 6 – 9 June 2010, Brugge, Belgium, 455–458.

[3] M.Nakagawa, Promoting discussion in Safety Review Activity, Chemical Engineering Transactions, Vol. 31, 2013

[4] Lees F.P., 1996, Loss prevention in the process industries second edition, Butterworth-Heinemann, Oxford, UK, Vol.1, 9/18.

[5] Center for Chemical Process Safety (CCPS), 2000, Guidelines for Chemical Process Quantitative Risk Analysis Second Edition. American Institute of Chemical Engineers, New York, NY.

[6] Center for Chemical Process Safety (CCPS), 1994, Guidelines for Hazard Evaluation Procedures. American Institute of Chemical Engineers, New York, NY.

[7] Center for Chemical Process Safety (CCPS), 1993, Guidelines for Safe Automation of Chemical Processes. American Institute of Chemical,Engineers,NewYork.

Abstract

This paper deals with the subject of intentional competency development which is necessary for establishing and sustaining process safety management in the industry. Why competency is important, frame work of competency development system, some methodologies which can help in reaching proficient level of competency are discussed. Also it briefly examines common gaps and factors affecting reliability of execution.

Introduction

Intentional competency is the ability to do something successfully and /or efficiently. It also means a cluster of related abilities, commitments, knowledge, and skills that enable a person (or an organization) to act effectively in a job or situation. In process safety reference it means developing competent organization and work force which understands the requirements of process safety management and has the ability to predict the consequences of their actions and inactions, make necessary plans/systems /procedures and can execute the same with required perfection.

All of us are aware that today process industry has much better designed plants, SIS (safety instrumented systems) controls have improved a lot over the years. Moreover equipment reliability and maintenance systems have also come a long way through Reliability engineering and specific PSM systems are in place.

Still process incidents having significant environment, people and legal consequences keep on happening. One reason which is cited in all incident investigations is "competency" which brings both individual as well as organization capability in focus.

Author: Pravesh Agrawal, Reliance Industries Limited, POY plant, Hazira

In this article we will deliberate on following topics related to competency development and management from process safety perspective:

- Why competency is important?
- Frame work and some tools of a competency development /management system.
- Common gaps and ensuring reliability of execution.

Why competency is important

Any organization has a responsibility to ensure the safety of its workers, the environment and also the public at large. Thus organization has to ensure that their leadership, staff and their contractors are fully competent in proactively managing process safety and responding to major accident hazards .In addition competency adds to the bottom line of the organization as it leads to reliable and efficient operation.

Similarly for individuals, competency insures their personal safety as well as safety of their team members. Apart from this, competency assurance gives individuals a sense of "doing it right" which is a great morale booster. It gives employees a sense of achievement, and purpose which in itself enhances their job satisfaction.

Frame work and some tools of a competency development /management system

Frame work for a competency development and management system can be broadly classified into following sub sections:

- **Management commitment**

Like any other major industry initiative management commitment is the indispensable constituent for competency development. Committing funds and announcing the policies is relatively easier part, but difficult part is the persistence, zeal, efforts & patience required to see the commitment through.

It also requires identifying leaders who believe in the competency development and have necessary people skills to implement the same .Technical skills to make necessary SOP's /SMPs /TBRA / PHA Specifications are easier to come by but people skill and execution of the plan are the areas where most companies falter .

Management's primary role is to develop a learning culture which promotes competency. This requires a consistent effort by the leadership & also it takes quite extended time.

Author: Pravesh Agrawal, Reliance Industries Limited, POY plant, Hazira

Expediting the culture developing effort has to be done in a very careful manner otherwise it may results into wastage of resources & efforts .It is good to have ambitious targets but more necessary is that targets should be achievable. Steady progress, may be on the slower side, is the only way to achieve reliable implementation. It is also important that initial implementation strategy is adopted after proper evaluation and it should be commensurate with the working methodology of the organization. Too many course corrections confuse operating staff and seriousness of the competency developing efforts is lost on them.

➢ **Training**

Training is at the core of competency development. Purpose of the training should be to provide all engineering and operating personnel with the appropriate procedure , information, guidelines , tools, and understanding to conduct tasks in alignment with the process safety requirements . Training program should include:

- Initial over view training.
- Basic skill training.
- Job task training with emphasis on safety critical task
- Statutory requirements of the site and organization.
- Emergency response and control.
- Refresher and supplemental training
- Recordkeeping.
- Training of instructors.
- Mandatory Training requirements for any job position.

Training modes may include a combination of class room training, field training, skill demonstration or simulator training for complex automated processes.

➢ **MRC**

Concept of Minimum required Competency (MRC) is one of the tools to ascertain the training needs for any job role. Experienced persons should frame the MRC document for all job roles which require PSM compliance. All training efforts can be further detailed based upon MRC.

While defining MRC specific process safety needs of the site/plant should necessarily be taken into account so individual gets the relevant training.

- **Assessment**

Training must include assessment, which may be a combination of written tests, skill demonstration, wherever appropriate, and final assessment by the line organization that a person is competent to independently complete a given task/job in a safe manner.

- ✓ **Case study – Training , Skill demonstration & Assessment**

In an organization pump bearing failures were on higher side leading to very low MTBF. After analysis, it emerged that one of the reason was improper alignment between pump & motor. Further analysis revealed that doing alignment using conventional method of dial gauges was time consuming & requires skilled mechanics which were scarce. Organization decided to go for laser alignment tool kit which facilitates precise and swift alignment. But pump failures, on the contrary increased after laser alignment kit was put into use. What had actually happened was that proper training, skill demonstration & assessment was not done , thus mechanics were not competent to use the new technology leading to even poor quality of the alignment.

- **Fit for duty**

Another area of concern while considering the training part is to ensure that personnel handling hazardous materials and activities are fit for duty and are not compromised by external influences e.g Alcohol, lack of sleep, illness or other personal issues. Trained and motivated supervisors, A good HR program, and alert team members can accomplish this task.

- **Contractors Training:**

How much training resource should an organization spend on training of contractors? Now a days with trend of reducing company employees & out sourcing of more and more jobs, it is imperative that contractor have as much knowledge about process safety as company employees.

In case of residential contractors it is much more important as they work on safety critical task on regular basis. . Thus all out sourced personnel performing safety critical tasks should go through all the hazard related training as identified for the company employee of the similar position. Organization should ensure this through regular assessment and audits. Safety performance of the contractors should be an important criterion for renewal of the contracts.

- **Some tools for competency development /management**

Over the years various methods/tools/systems have emerged which can help organizations achieve greater competency levels for managing process safety. Some of those are listed below:

- TBRA (Task based risk assessments) for mission critical tasks.
- LCI (last Chance Inspection) for mechanics.
- MERI (Minimum Essential Required Inspection) check list for spares identified as PSM critical.
- PDCA / JCC (Job Cycle check) for SOP and SMP reviews.
- Risk based inspections (RBI) , FMEA /RCM for reliability improvement
- PHA (Process hazard Analysis) for risk assessment and mitigation.
- Simulations for complex ERCP scenarios.
- LLF (Look, Listen and Feel) check lists for operators/mechanics rounds.

Use of these methodologies involve all levels of company/contractor employees and a feeling of owner ship and team spirit starts building , which is another needed element for process safety. Involving all persons also bridges the gap between what is written in procedures and what is executed in the field leading to better compliance.

✓ **Case study – LCI (Last Chance Inspection)**

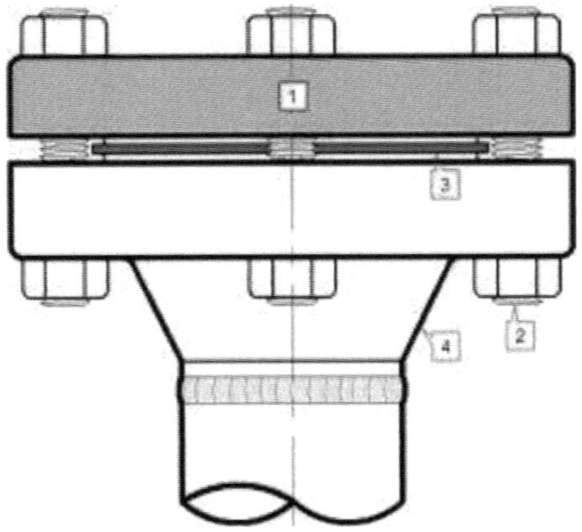

In a petrochemical company, shut down was in final stages and 24" flange of a vessel handling chlorine was being boxed up. Job was to be completed in night shift & maintenance crew took out the gasket from its packing for installation. Before they could install the same, due to some emergency, maintenance crew was shifted to another activity and job could not be completed in night shift. Day shift crew took up the job and last chance inspection just before the installation revealed that gasket was in damaged condition. Culture of the organization and competence of the maintenance crew in LCI saved the day for the company.

Common gaps and ensuring reliability of execution

Organizations can set up systems and procedures to develop process safety competency of the individuals but how do they ensure and motivate persons to follow these on continuous basis. Following are some the common gaps and factors affecting reliability of execution:

Consistent implementation of procedure and systems: Leaders should believe in the systems and procedures which they are propagating and should make available sufficient resources to implement the same .Developing culture is the key .Employees have great observation power, like a child, and imbibe and develop their competency in discretionary ways on procedures / systems which are followed on regular basis i.e they follow the cuture. Procedures /systems which are followed on selective basis will see maximum violations, as employees feel it is not necessary to build competency on these.

Differentiating between PSM critical &others: This point is made many a times but generally overlooked into implementation. It is very important that a clear differentiation is made between PSM critical equipment's /activities/procedures and others. This will help in building additional competencies required for managing process safety critical task.

Continuous review of the trainings being imparted: Organizations impart a huge amount of training to their employees now a days. It is necessary that only "hours of training" are not monitored but how useful it is to the employee and organization is also looked into on

continuous basis. Taking feedback from operating staff and redesigning training courses is the key for continual improvement.

Awarding Competence: It is a must that individual /teams which are excelling in "competency" area should be recognized in common forums of the organization. How do we judge the "competency"? Organizations should device their own ways e.g. monitoring process safety KPI's i.e spills , PSV popping , containment controls , auditing the SOP and actual work done in the field , interviewing mechanics on a hazardous maintenance work , compliance to key learnings of the incident or near miss etc. .

Promoting Innovation : As John Maynard Keynes famously said ""The difficulty lies not so much in developing new ideas as in escaping from old ones." , it is very important that new ideas , out of box thinking is listened to in the organization , to promote innovation in developing competency .

 # Minimize use of disciplinary actions: After spending time and resources on process safety initiatives when required results are not obtained, organizations start using disciplinary actions for minor violations also. This generally does not help. In most of the cases , fact remains that time & resources were not properly spent , ground realities were not taken into account , methodology of implementation was not correct , procedures are unnecessarily cumbersome, sufficient resources were not provided. Before initiating any disciplinary actions it is necessary to look at the present culture of doing that activity in his/her area. Disciplinary actions should be limited to intentional, habituated, repetitive offenders. Use of frequent disciplinary actions just to make an example, creates an environment of fear which impedes learning.

HR practices: Last but not the least, good HR practices are essential to create an environment of learning which leads to competent employees. "Good" does not necessarily mean higher wages. More important are clear job roles, responsibilities and authorities, defined and consistent policies, transparent systems, fair evaluations, clear & continuous communication and growth avenues.

Conclusion

Competency development is essential to ensure process safety in the modern process industry. Proficient level of competency can be achieved by management commitment, doing need assessment, Framing organization specific training programs, continuous monitoring, imbibing modern competency development tools and continual improvement.

References:

Images used are from open internet using Google image search.

Author: Pravesh Agrawal, Reliance Industries Limited, POY plant, Hazira

Process Safety Implementation Challenges in Cross Cultural Environment

Mahesh Murthy
Process Safety Manager - Maersk Oil Kazakhstan GmbH
Mahesh.murthy@outlook.com; +372 5432 9002
CCPS – Global Process Safety Summit
Mumbai – India
December 2014

Engineers do not fail in engineering ....
but fail in imagination

Rex Tillerson, CEO – ExxonMobil – Congressional Committee – US

What am I talking today

- Setting the context
- Concept of safety lifecycle
- Process Safety – theory versus practical
- Concluding remarks

Severity Rating	Consequences							Likelihood					
								A	B	C	D	E	F
								$< 10^{-6}$	$\geq 10^{-6}$ to $< 10^{-4}$	$\geq 10^{-4}$ to $< 10^{-3}$	$\geq 10^{-3}$ to $< 10^{-2}$	$\geq 10^{-2}$ to $< 10^{-1}$	$\geq 10^{-1}$
	People	Business / Assets	Environment	Reputation	Security	Economic (USD)	Schedule	Remote	Extremely Unlikely	Very Unlikely	Unlikely	Forseeable	Likely
1	Minor injury. Minimal health effect	Minimal damage	Minimal effect	Minimal impact	Minimal effect	Minimal loss	Minimal delay						
2	Injury requiring medical treatment. Short term health effect	Slight damage	Slight effect	Slight impact	Slight effect	Slight loss	Slight delay						
3	Lost workday injury. Medium term health effect	Minor damage	Minor effect	Limited impact	Minor effect	Minor loss	Minor delay						
4	Major injury. Long-term disability / health effect	Localised damage	Localised effect	Considerable impact	Localised effect	Moderate loss	Moderate delay						
5	Single fatality or multiple major injuries with long term effects	Major damage	Major effect	National impact	Major effect	Major loss	Major delay						
6	Multiple fatalities	Extensive damage	Massive effect	Major international impact	Massive effect	Massive loss	Massive delay						

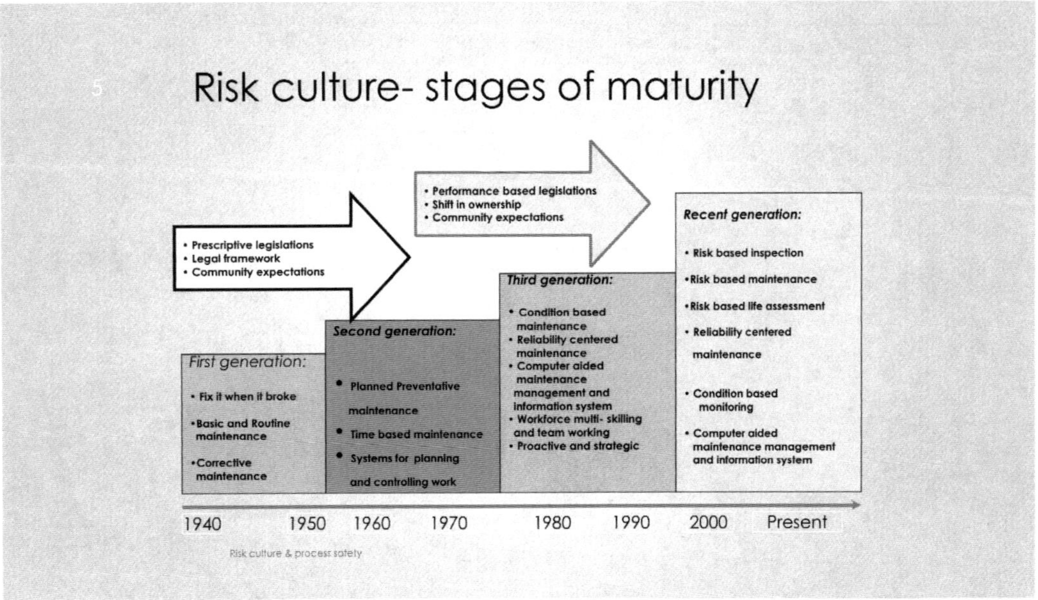

If you think safety is expensive……try having an accident – Trevor Kletz

Some useless investigations/audit recommendations..................

- *Safety culture did not exist................*
- *Contractor Management not fully established*
- *Lack of adequate staff training and competency*
- *Risk was not fully understood*
- *Previous audit recommendations were not implemented*
- *Consider implementing risk based asset management*
- *Site was not fully compliant with local regulatory requirements*

When did you last hear someone say......

- *LPG bullets need to be relocated to meet the minimum separation distance in accordance with API 2510*
- *Pipework need to be upgraded to impact tested stainless steel to avoid brittle failure due to process conditions*
- *HAZOP actions not closed in time as per company risk management policy*

......

- *PSV does not comply with API 521 requirements for sizing*
- *ESD valve does not meet the performance standard to meet SIL 3 requirement*

So what is process safety ?

Process Safety is a blend of engineering and management skills focused on preventing catastrophic accidents, particularly explosions, fires, and toxic releases, associated with the use of chemicals and petroleum products – CCPS

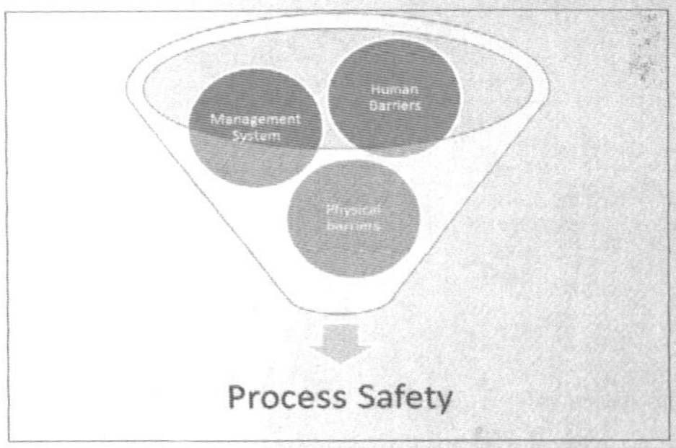

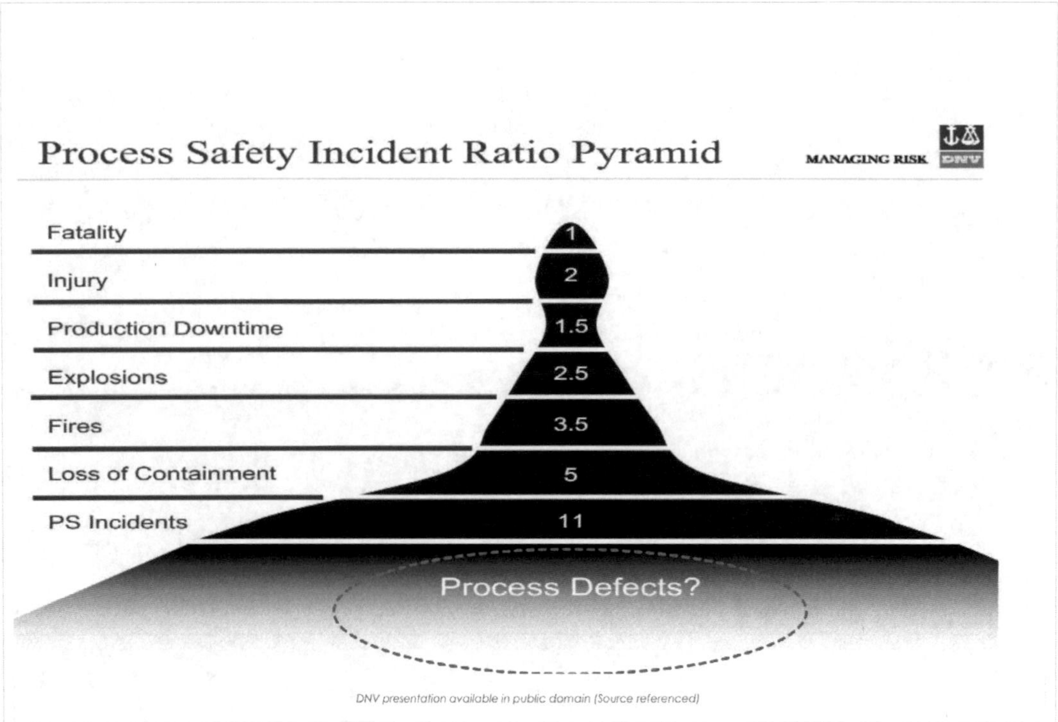
Process Safety Incident Ratio Pyramid
MANAGING RISK
DNV
Fatality
1
Injury
2
Production Downtime
1.5
Explosions
2.5
Fires
3.5
Loss of Containment
5
PS Incidents
11
Process Defects?
DNV presentation available in public domain (Source referenced)

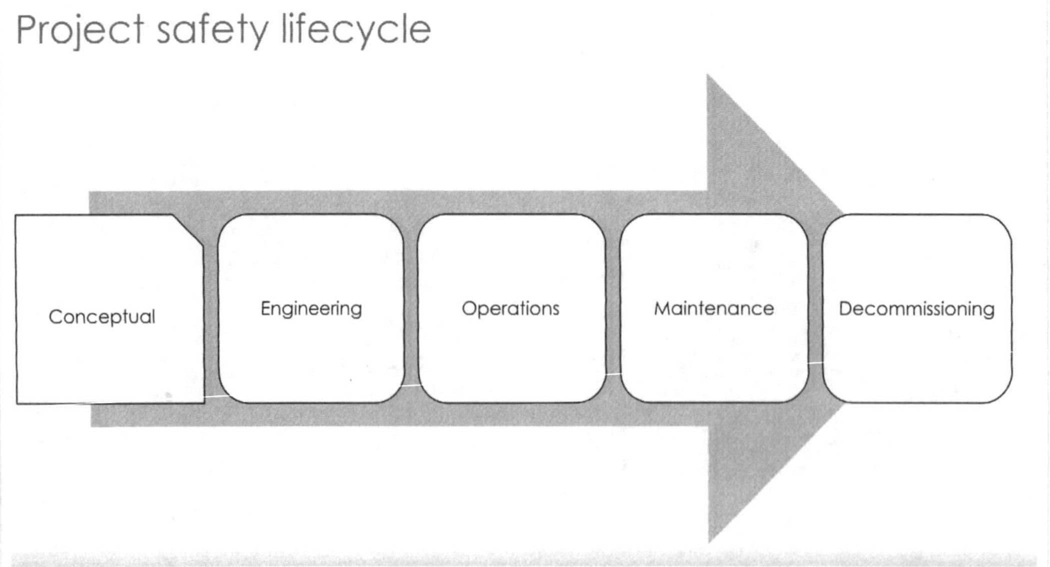
Project safety lifecycle
Conceptual
Engineering
Operations
Maintenance
Decommissioning

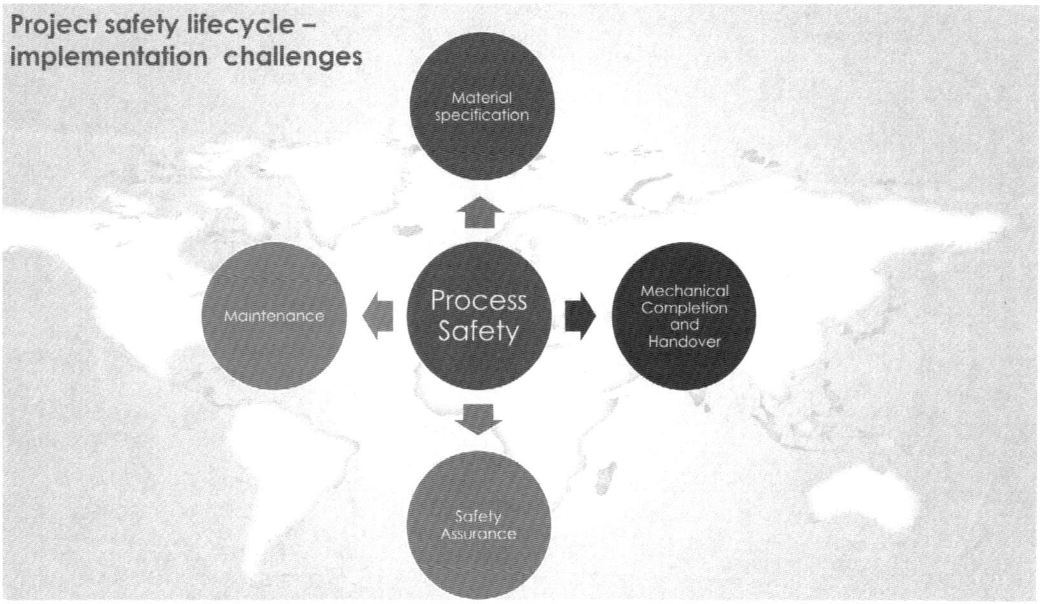

Material specification

- Estimated 50% of 100 major hydrocarbon losses to date have root cause of improper material specification
- 60% of Instrumentation failures have resulted in hydrocarbon losses due to improper material specification
- Heat treatment deficiencies due to cheap imports have resulted in increased pressure vessels failure rate
- New materials like GRE/Composites have not been time tested with minimal Quality Assurance applied

Safety assurance

- Safe Operating Limits
- Causal factors for procedural violation should be understood
- Procedural streamlining should be as important as having a plan for re-HAZOP or corporate planning
- Human Factors expert need to comment on the layout of the procedures
- Audit scope of works need to be specific and against clear criteria

Interesting facts

- Around 85% of Process safety investigations found violations of procedures
- Plant managers have less than 4 minutes of their time to review and approve a procedure
- Safety video messages longer than 4 minutes tend to have very little impact on workforce
- Executive with soiled plant PPE commands more respect than well dressed executives

Mechanical completion and handover phase

Challenges in procuring latest process safety information;

- Flare and relief study reports
- PSV narratives
- Performance Standard for ESD valves
- SIL dossiers and calculations
- Vent dispersion calculations
- HYSYS, FlareNet and Phast Risk native files
- Vendor packages design basis

Project focus is on cost and schedule, not on documentation !!

Maintenance

Have you established;

- Safety Critical Element (SCE) register
- Work order to leak ratio
- Cumulative risk of SCE deferrals instead of individual risk
- Mean Time to Failure of SCE
- Corrosion loops for all plant areas

Interesting fact

- PSV maintenance late by due date by 10% is a non-conformance !!

Due date is inclusive of 10% allowance !

Concluding remarks

- Process Safety begins at the conceptual stage
- Design changes should not be ruled out to achieve process safety at any stage of lifecycle
- Detailed planning on organisation manning levels at all project safety lifecycle stages
- Human Factors in seeking meaningful safety assurances

Common sense is the most uncommon thing these days...do not let this happen to you

AcuTech
PROCESS RISK MANAGEMENT

US Executive Order for Improving Chemical Safety and Security

Prepared for Presentation at
CCPS Global Summit on Process Safety
December 15-16, 2014
Lalit Hotel, Mumbai, India

AcuTech Group, Inc.
Washington DC | Houston | Philadelphia | San Francisco
Dubai | Shanghai
www.acutech-consulting.com

AcuTech
PROCESS RISK MANAGEMENT

Outline

- Need for improved PSM performance
- US Regulatory Update
- Conclusions and Q&A

AcuTech
PROCESS RISK MANAGEMENT

Process Safety Management (PSM)

- Process Safety Management (PSM) is a "blend of engineering and management skills focused on preventing catastrophic accidents, particularly explosions, fire and toxic releases associated with the use of chemicals and petroleum products" (AIChE, CCPS, 2010)

AcuTech
PROCESS RISK MANAGEMENT

Challenge - Improvement of Performance in the Management of Risk

- Requires **good corporate governance**
- This requires the establishment of:
  - Objectives and targets
  - Metrics to monitor and review performance
  - Audits and management reviews
  - The right culture facilitated by:
    - collaborative efforts to learn, share and address issues
    - an operating discipline
    - A mindset of improvement and Operational Excellence

How Are We Doing?

- Perception that improvements in the regulatory approach are required since little change from the origins in 1992.
- The President issued Executive Order (EO) on 1 Aug 2013 directing six federal agencies to evaluate and improve the safety and security of the U.S. chemical industry.
- The EO was issued in the wake of the accident at an ammonium nitrate distribution facility in West, TX in April 2013.
- In May 2014 the Working Group of these six federal agencies issued an interim report to the President on their preliminary conclusions.
- In December 2014 there was a US Senate hearing on the matter giving more insight to the expected outcome and timeline

5

US Refinery National Emphasis Program Year 1 Results

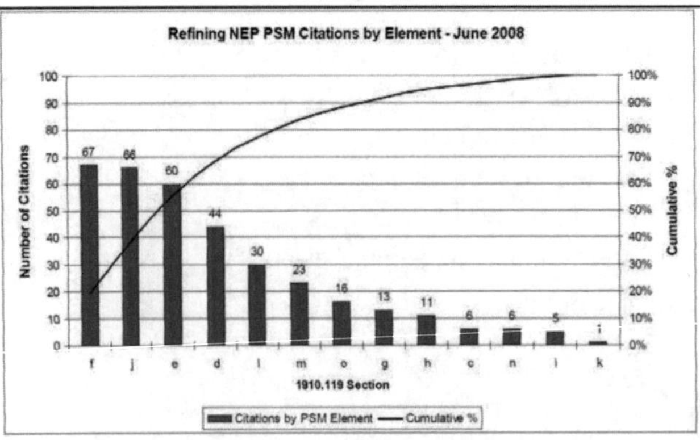

Figure 1. Number of Citations and Cumulative Percentages, By PSM Element

Refinery NEP Year 1 Results

Paragraph	PSM Program Element	Paragraph	PSM Program Element
c	Employee Participation	j	Mechanical Integrity
d	Process Safety Information	k	Hot Work Permit
e	Process Hazard Analysis	l	Management of Change
f	Operating Procedures	m	Incident Investigation
g	Training	n	Emergency Planning & Response
h	Contractors	o	Compliance Audits
i	Pre-startup Review	p	Trade Secrets

Reports on BP Texas City

West Fertilizer Explosion and Fire
April 17, 2013
West, Texas

- Detonation of ammonium nitrate
- 15 fatalities – 12 first responders
- 160 were injured and more than 150 buildings were damaged or destroyed

West - Before

AcuTech
PROCESS RISK MANAGEMENT

West - After

11

AcuTech
PROCESS RISK MANAGEMENT

West Fertilizer Explosion and Fire

- The West Volunteer Fire Department did not have an emergency plan for a fire at the plant
- The department's dozens of volunteer firefighters did not know the explosive potential of ammonium nitrate.
- West Fertilizer Co. never reported the ammonium nitrate to the Department of Homeland Security, though it stored more than the 400 pounds of the chemical. That's the threshold that triggers self-reporting.
- DHS did not have West on its radar of dangerous chemical sites even though the company reported the chemicals to the McLennan County emergency planning committee and the state health department.

12

West Fertilizer Explosion and Fire – Lessons Learned

AcuTech
PROCESS RISK MANAGEMENT

- Industrial sites handling, storing, or manufacturing hazardous chemicals must be properly managed by government including accurate inventory, assessing safety and security plans, ensuring the public risk is tolerable, and enforcing existing regulations
- Industry has to be responsible to understanding their hazards, reporting the presence of hazardous chemicals and to reducing risks to a minimum through process safety management programs
- Industry and local or national emergency response agencies need to be aware of the hazards and to have realistic plans for managing potential impacts of incidents

13

Executive Order 13650 of August 1, 2013

Improving Chemical Facility Safety and Security

AcuTech
PROCESS RISK MANAGEMENT

Federal Register / Vol. 78, No. 152 / Wednesday, August 7, 2013 / Presidential Documents **48029**

Presidential Documents

Executive Order 13650 of August 1, 2013

Improving Chemical Facility Safety and Security

By the authority vested in me as President by the Constitution and the laws of the United States of America, it is hereby ordered as follows:

Section 1. *Purpose.* Chemicals, and the facilities where they are manufactured, stored, distributed, and used, are essential to today's economy. Past and recent tragedies have reminded us, however, that the handling and storage of chemicals are not without risk. The Federal Government has developed and implemented numerous programs aimed at reducing the safety risks and security risks associated with hazardous chemicals. However, additional measures can be taken by executive departments and agencies (agencies)

14

Executive Order 13650 of August 1, 2013

Improving Chemical Facility Safety and Security

- President Obama signed an executive order "Improving Chemical Facility Safety and Security."
- The Order has the potential to broadly impact industrial facilities that use, manufacture, store or handle high quantities of hazardous chemicals.
- Key aspects:
  - enhancing information sharing between government critical infrastructure owners;
  - addressing policy, regulation and standards modernization;
  - and identifying best practices.

Outlook

- Greater cooperation and information sharing between federal agencies
- Increased Government outreach to state and local actors
- Greater efforts to identify noncompliant chemical facilities
- More inspections to enforce regulations
- Legislative activity for modifying regulations
- Increased safety and security scrutiny (Note: the US chemical security regulation was extended by Congress for 4 more years in a new bill and will be modernized as well)

Introduction

- The EO Working Group report addresses the following areas:
  - Strengthening Community Planning and Preparedness
  - Enhancing Federal Operational Coordination
  - Improving Data Management
  - ***Modernizing Policies and Regulations***
  - Incorporating Stakeholder Feedback and Developing Best Practices
- The EO Working Group report outlines a Federal Action Plan to address these areas.

Introduction

- Within the topic of *Modernizing Policies and Regulations* the EO Working Group report addresses the following issues/topics:
  - **Modernizing OSHA's PSM Standard to Improve Safety and Enforcement**
  - **Modernizing EPA's RMP Regulation**
  - Enhancing Ammonium Nitrate Safety and Security
  - **Promoting Safer Technology and Alternatives**
  - Building a Stronger CFATS Program
  - Developing Guidance and Outreach Programs to Help Work with States to Improve Safe Drinking Water Act (SDWA) Measures to Prevent and
  - Prepare for Chemical Spills
  - **Increasing OSHA Penalties**
  - Pursuing Statutory Amendment to the Safe Explosives Act (SEA)
  - Improving Process for Notification of Stored Explosives to Fire Authorities

Modernization of the PSM Standard

- The Federal Action Plan recommends the following modernizations of OSHA's PSM Standard:
  - Revise the current interpretation of *retail facilities* to more accurately reflect the original intent of the exemption as expressed in the PSM Preamble.
  - Revise the current interpretation of *chemical concentrations* covered by OSHA's PSM standard to more clearly describe what is covered and align with better established practices.
- Both of these issues were addressed in OSHA's RFI of December 9, 2013.

19

Modernization of the PSM Standard

- Based on information collected from the OSHA RFI and the EO Section 6 Options document, the PSM rulemaking will consider the following revisions (among others):
  - *Clarifying the PSM Standard to incorporate lessons learned from enforcement, incident investigation, and advancements in industry practices, root cause analysis, process safety metrics, enhanced employee involvement, third-party audits, and emergency response practices.*
  - Addressing ammonium nitrate hazards through one or both of the following options:
    - Covering reactive chemical hazards under the PSM.
    - Adding ammonium nitrate specifically to the PSM Appendix A highly hazardous chemicals list.

20

Modernization of the PSM Standard

- Cont'd:
  - Adding substances or classes of substances to the PSM Appendix A List of Highly Hazardous Chemicals and providing more expedient methods for future updates.
  - Expanding coverage and requirements for reactive chemical hazards, which have resulted in many incidents.
  - Covering oil and gas drilling and servicing operations that currently are exempt from PSM coverage.
  - *Continuing harmonization with EPA's RMP regulation.*

21

Modernization of the PSM Standard

- Cont'd:
  - ***Requiring analysis of safer technology and alternatives.***
  - Requiring coordination between chemical facilities and emergency responders to ensure that emergency responders know how to use chemical information to safely respond to accidental releases, possibly including exercises and drills.
- Several of these issues/topics were explicitly addressed in OSHA's RFI of December 9, 2013.

22

Modernization of the RMP Rule

- The Federal Action Plan recommends the following to modernize the RMP Rule:
  - *Revising mechanical integrity requirements* of safety-related equipment. The Working Group report does not describe what changes EPA should make, but these are likely to be the same issues addressed in the OSHA RFI on MI, i.e., definition of a RAGAGEP.
  - Adding new requirements for *automated detection and monitoring systems*, or adding performance measures for facilities already using these systems that would supplement the existing process hazard analysis (PHA) and/or emergency response requirements.
  - Establishing an obligation to track and conduct *root cause analyses* of frequent process events and near misses.
  - EPA to issue RFI for RMP Rule in Summer 2014.

23

Modernization of the RMP Rule

- Cont'd:
  - Requiring employees to implement a *stop work authority* for employees who witness an activity that creates a threat of danger and providing clearly defined requirements to establish an *ultimate authority* on the facility for operational safety and decision making.
  - Strengthening contractor safety requirements.
  - Establishing mechanisms to implement the newest available technologies and methods being brought to bear in chemical risk management, PHA, and emergency response.
  - Requiring compliance audits be done by an *independent auditor* to increase the rigor and objectivity of the audit.
  - Clarifying what is required in order for a *PHA to be updated and revalidated*, requiring revalidating PHAs more frequently than every 5 years, and requiring certain events such as an incident to trigger PHA revalidations prior to the next scheduled 5-year revalidation.

24

AcuTech PROCESS RISK MANAGEMENT

Modernization of the RMP Rule

- Cont'd:

  Establishing new performance measurement and management review requirements, such as:

  – A measurements and metrics requirement.

  – A management review and continuous improvement requirement.

  – A process safety competency requirement to focus on organizational learning.

AcuTech PROCESS RISK MANAGEMENT

Modernization of the RMP Rule

- Cont'd:
  – Clarifying emergency planning requirements to ensure effective coordination with community responders and ensuring facility personnel practice the plans, and that the drills and exercises include offsite responders if possible.
  – Enhancing disclosure of key elements of a facility's risk management plan and program from facilities to improve community understanding of chemicals.
  – Incorporating examination of the use of safer technology alternatives into the PHA.
  – Using the Acute Exposure Guideline Levels (AEGLs) to recalculate RMP reporting thresholds and toxic endpoints for offsite consequence analyses.

Modernization of the RMP Rule

- Cont'd:
  - Facility and equipment siting factors and 'buffer zones' between the facility fence line and public receptors
  - Whether the current worst-case scenario, which involves a catastrophic failure of the single largest vessel, should also assess the total catastrophic failure of multiple smaller vessels stored in close proximity to one another.

Modernization of the RMP Rule

- Timeline:
  - EPA to propose amendments to the RMP Rule by Summer 2015.
  - EPA to finalize revised RMP Rule by Summer 2016.
  - These recommendations to modernize the RMP are lengthier than the list for PSM modernization and are expected to be more intensive. EPA testified last week in a Senate committee that they expected to have a new regulation in place "prior to the end of the current administration".

AcuTech's Approach to Process Safety Excellence

- AcuTech follows a classic Deming Plan-Do-Check-Act approach for the development, implementation and management of multiple tasks and disciplines to manage risk:
  - **Plan**
    - Documenting the management system
    - Defining requirements and objectives
    - Regulatory, Internal, Industry Best Practices
  - **Do**
    - Implementation of the system
  - **Check**
    - Audits, risk assessments, internal review
    - Checklists
    - Key performance indicators review
  - **Act**
    - Management review
    - Setting of improvement objectives
    - Monitoring progress on implementation or action items

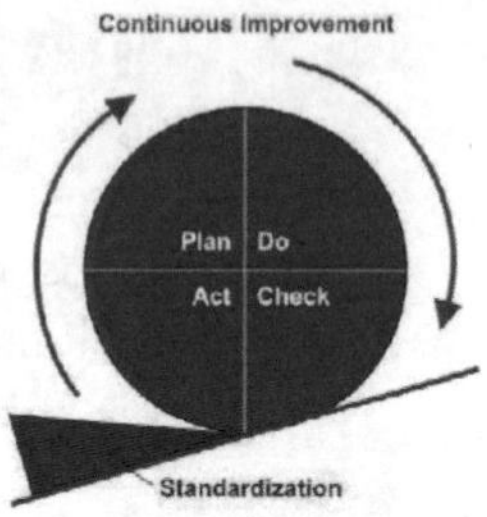

Conclusions

- The following overall conclusions are relevant for facilities covered by PSM and RMP:
  - The items in the OSHA RFI have not been deleted or diluted. The EO Working Group Report strengthens them.
  - RMP modernization could be even more extensive than PSM modernization.
  - IST is now part of the planned PSM and RMP modernization plan via inclusion as an analysis (PSM) or via examination in the PHA element (RMP).
  - All of the recommendations are short term (one year) actions, except for one in Safer Technology and Alternatives.

AcuTech Consulting Group
1919 Gallows Road
Suite 900
Vienna, VA 22182 USA
www.acutech-consulting.com

About the Presenter

David Moore is the President and CEO of the AcuTech Consulting Group, a risk management consulting firm based in Vienna, Virginia, and founded in 1994 (www.acutech-consulting.com). AcuTech has been a sponsor of CCPS since inception and serves on the Technical Steering Committee. AcuTech has developed four guidelines for CCPS (Inherent Safety, Metrics, Auditing, and Security Risk Analysis).

Mr. Moore has over 34 years of experience in chemical safety and security management and is a recognized expert in and frequent speaker on these topics. He has provided risk consulting services and training to industrial companies worldwide, including at petroleum refineries, chemical plants, gas plants, pipelines, pharmaceuticals, defense, biotech, and manufacturing plants. Mr. Moore has taught process safety and security courses to many of the world's largest corporations and the US Government. He has been a consultant to USOSHA, DHS, USCG, AIChE, API, USEPA, and the American Chemistry Council.

Mr. Moore was formerly a Senior Engineer with Mobil Corporation and a Fire Protection Engineer with the National Fire Protection Association . Mr. Moore is a Registered Professional Engineer. He has an MBA, (NYU 1987), and a B.Sc., Fire Protection Engineering (University of Md. 1979).

30 years on from Bhopal

Charles Cowley – CCPS Staff Consultant

Scott Berger - CCPS Executive Director

2014 CCPS Global Summit on Process Safety
The Lalit, Mumbai, India
December 15-16, 2014

30 years on from Bhopal

1. The situation in Bhopal today
2. Views of experts
3. **Call for Action** – What needs to be done?

The situation in Bhopal today

Three key aspects

- Condition of the old Union Carbide site
- Soil and groundwater contamination
- Public health and social situation

Condition of the old Union Carbide site

- Stored chemicals
- On-site soil contamination (mercury… others)
- Old buildings and equipment

Bags of formulated Sevin still left in the plant - Metal storage building corroded; now admits rainwater which carries Sevin into surface and groundwater

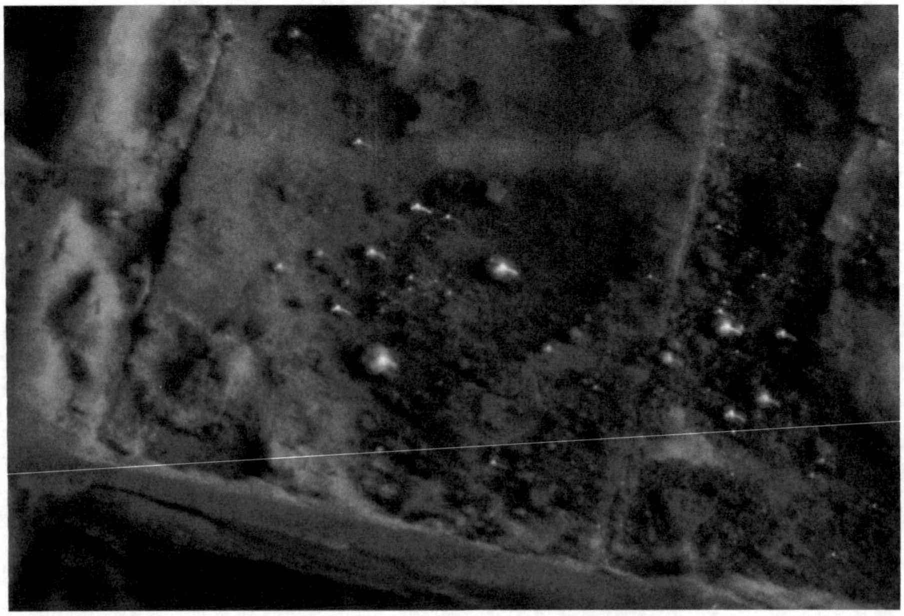

Mercury. Had been used in a liquid seal. Metal corroded and mercury now on ground

A tank which contained an unknown process residue.
Tank corroded and residue now sits on ground - 10 years later, residue may have washed away

The scrubber which was out of service

The flare which was by-passed

The refrigeration system which was out of service

Filters which according to one theory were being incorrectly washed,
leading to water contamination of the MIC. White material is asbestos, not snow

Papers on the floor are permits-to-work which had been stored in the control room

Sign on control panel.
Also posted on control panel were instructions for responding to an MIC leak, which were not put in place until hours after the leak started

An AIChE Technology Alliance
CCPS
Center for Chemical Process Safety

Soil and groundwater contamination

- Solar Evaporation Pond
- Groundwater

Views of experts

- **Medical:** Professor Paul Cullinan

  Professor of Occupational and Environmental Respiratory Disease, National Heart & Lung Institute - Imperial College, London; and Member of International Medical Commission on Bhopal

- **Environmental:** Centre for Science & Environment, India

- **Social:** Syed Tabish Ali

  Chingari Trust – non-profit, non-political charitable trust, providing rehabilitation support to the Bhopal gas-affected community – mainly children and their mothers

extracted from IChemE LPB Special Edition on Bhopal

Medical: Professor Paul Cullinan

What we know of MIC (above) suggests that residual damage to the lower airways in a proportion of those exposed would be expected and there is reasonable evidence that this is so, at least in those who were adults at the time of the disaster. This damage is manifest as a fixed scarring of the small airways ('obliterative bronchiolitis') which causes those affected to be more or less breathless on exertion or even at rest; it is not amenable to drug treatment.

several community-based organisations (notably the Chingari and Sambhavna clinics) where large numbers of people from the most marginalised and needy sectors of the city are offered care in an environment of extraordinary tenderness. While there is no proof that the conditions being treated in these settings are a direct result of the 1984 accident, or from ongoing contamination from abandoned pesticide, there is no proof that they are not. In a story that has few heroes, those who tirelessly work there are worthy of our support.

extracted from IChemE LPB Special Edition on Bhopal

Environmental:
Centre for Science & Environment, India

Immediate measures

1) Securing the site and SEP area by fencing and guarding to prevent access of people, especially children, hence their exposure to toxic chemicals; stopping construction in the SEP area; and protecting annual surface water runoff from the site during monsoon.
2) Excavation and recovery of all the waste from the site; Characterisation and inventorisation of the collected waste for proper treatment and disposal.
3) Characterisation of the 350 tonne wastes that is stored at the site and the results to be shared in the public domain. Under the supervision of the CPCB and affected community, incinerable waste is to be incinerated after the stabilisation of the trial results at Pithampur.

extracted from IChemE LPB Special Edition on Bhopal

Environmental: Centre for Science & Environment, India

Medium and long-term measures

1) Groundwater contamination assessment through detailed field investigation and laboratory analysis to develop a remediation plan. Possibility of hydraulic containment is to be explored as an interim containment measure.
2) Characterisation and remediation of the waste dumped in SEP area, particularly the landfill, to prevent continued contamination of the groundwater in the local area.
3) Detoxification, dismantling and decommissioning of the plant after preserving structures such as MIC plant including the vent, vent scrubber, storage tanks and control room.
4) Remediation of the UCIL site that involves building a memorial and centre of excellence for industrial disaster management after decontaminating the site.

Social: Syed Tabish Ali Chingari Trust

At present there are more than 600 such special children registered with Chingari Trust but because of resource limitations, we are only able to treat 180 children regularly, providing them with physiotherapy, speech therapy, special education, sports, mid-day meals, pick-up and drop-off facility, occupational therapy and other facilities free of cost. If these 180 children can show improvement then the other 420 children registered with us and many other such children born in gas and water affected families but not yet registered with Chingari Trust could also show improvement if provided with proper treatment, love and care.

www.chingaritrustbhopal.com

CHINGARI REHABILITATION CEN
(MANAGED BY CHINGARI TRUST)
WORKING FOR CHILDREN BORN WITH CONGENITAL DEFECTS TO VICTIMS OF
MANAGING TRUSTEES: MS. RASHEEDA BEE AND MS. CHAMPA DEVI SHUKLA

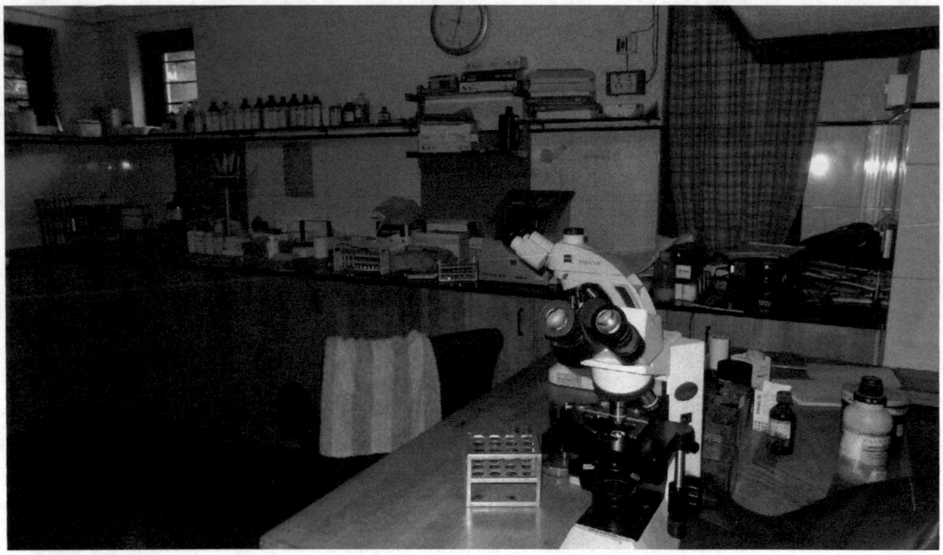

REMEMBER
BHOPAL
MUSEUM
Union Carbide Industrial Disaster 1984

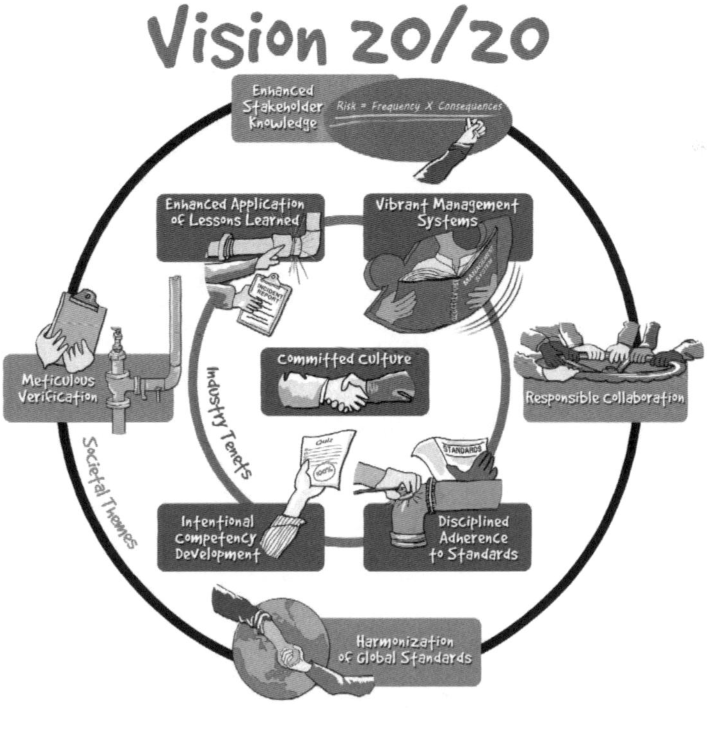
Vision 20/20
Enhanced Stakeholder Knowledge
Risk = Frequency X Consequences
Enhanced Application of Lessons Learned
Vibrant Management Systems
Committed Culture
Meticulous Verification
Responsible Collaboration
Industry Tenets
Societal Themes
Intentional Competency Development
Disciplined Adherence to Standards
STANDARDS
Harmonization of Global Standards

Responsible Collaboration

- Technical, government, labor, community, scientific, academic, and industrial organizations work together
- Remove barriers
- What can you do?

☑ Recognize the value that others bring to the table and commit to working together toward the common process safety goal.

☑ Be open-minded and focus on that common goal.

☑ Actively participate in or lead joint industry projects.

A Call To Action

...30 years is already too long

It's time to leverage our resources, knowledge and skills to:

- **Clean up the site and surroundings**
- **Support the local affected community**

www.chingaritrustbhopal.com Bhopal medical appeal

An AIChE Technology Alliance
CCPS
Center for Chemical Process Safety

Process Safety Leadership Forum

Hotel The Lalit
Mumbai, India
16th December, 2014
09:00 – 13:00

Facilitators:
Scott Berger – Executive Director, CCPS
Hirak Dutta – Executive Director, OISD
Louisa Nara – Technical Director, CCPS

"The Global Community Committed to Process Safety"

An AIChE Technology Alliance
CCPS
Center for Chemical Process Safety

Process Safety Leadership at the Board Level

"How I drive, and will drive,
Process Safety in my organization."

"The Global Community Committed to Process Safety"

An AIChE Technology Alliance
CCPS
Center for Chemical Process Safety

Summary Discussion Points

- Leadership Visibility – leadership must consciously engage and demonstrate a strong commitment.
- Leadership Development – need to have this ingrained in Schools of Management.
- Long Term Sustainability - profit driven companies need to understand Process Safety.
- Competency Assurance / Stewardship - Employees must be competent and confident and know expectations and individual roles. At all levels of employees must understand the consequences of their actions

"The Global Community Committed to Process Safety"

An AIChE Technology Alliance
CCPS
Center for Chemical Process Safety

Summary Discussion Points

- Link Process Safety and Operational Risk.
- Learn Best Practices from Other Industries.
- Share Incidents without the fear of legal implications.
- Understand **WHY** we are not Learning from Incidents (and address / correct).
- Align Contractors with company expectations and values.
- Focus on Partnership with industry, government, society.
- Management of Change and Management of Organization Change systems must be robust and must be implemented.

"The Global Community Committed to Process Safety"

Closing Thoughts

- Must always look at yourself first.
- EHS / PS must learn to talk the same talk as the CEO and other executives (10-day MBA).
- Enhance process safety implementation.
- Organization must have zero tolerance for violating life saving rules (e.g., entrance into confined spaces) and have a strong due diligence. Make expectations clear. Position in a care and positive way (I don't want to stand in front of your family if you violated a rule and put yourself at risk – with catastrophic consequence). An escalation matrix needs to be used in this process.

If you have any questions regarding this notes, please contact
Louisa Nara, CCPS Technical Director at louna@aiche.org

"The Global Community Committed to Process Safety"

AUTHOR INDEX